Data Structures using C

Data Structures using C
A Practical Approach for Beginners

Amol M Jagtap
Rajarambapu Institute of Technology, India
Ajit S Mali
Rajarambapu Institute of Technology, India

CRC Press is an imprint of the
Taylor & Francis Group, an **informa** business
A CHAPMAN & HALL BOOK

First edition published 2022
by CRC Press
6000 Broken Sound Parkway NW, Suite 300, Boca Raton, FL 33487-2742

and by CRC Press
2 Park Square, Milton Park, Abingdon, Oxon, OX14 4RN

CRC Press is an imprint of Taylor & Francis Group, LLC

Library of Congress Cataloging-in-Publication Data
Names: Jagtap, Amol M., author. | Mali, Ajit S., author.
Title: Data structures using C : a practical approach for beginners / Amol M. Jagtap, Rajarambapu Institute of Technology, India, Ajit S. Mali, Rajarambapu Institute of Technology, India.
Description: First edition. | London ; Boca Raton : C&H/CRC Press, 2022. | Includes bibliographical references and index.
Identifiers: LCCN 2021028843 (print) | LCCN 2021028844 (ebook) |
ISBN 9780367616311 (hbk) | ISBN 9780367616373 (pbk) | ISBN 9781003105800 (ebk)
Subjects: LCSH: Data structures (Computer science) | C (Computer program language)
Classification: LCC QA76.9.D35 J34 2022 (print) | LCC QA76.9.D35 (ebook) | DDC 005.7/3—dc23
LC record available at https://lccn.loc.gov/2021028843
LC ebook record available at https://lccn.loc.gov/2021028844

ISBN: 9780367616311 (hbk)
ISBN: 9780367616373 (pbk)
ISBN: 9781003105800 (ebk)

DOI: 10.1201/9781003105800

Typeset in Palatino
by codeMantra

Contents

Preface

It gives me great pleasure to come up with this book. This is motivated by the desire that we and others have had to further the evolution of the basic course in computer science.

This book is intended for teaching purposes. We believe it's more important for a practitioner to understand the principles required to select or design the data structure that will best solve a problem. Therefore, we designed this as a teaching text that covers most of the standard data structures. Some data structures that are not widely adopted are included as examples of important principles. Certain relatively new data structures that are expected to be widely used in the future are included.

One of the most important aspects of a course in data structures is that it is where students learn to program using pointers and dynamic memory allocation, by implementing data structures such as linked lists, stack, queue, trees and graph. In our curriculum, this is the first course where students make significant design because it often requires real data structures to motivate significant design exercises. Finally, the basic differences between access to data in memory and access to data on disk cannot be assessed without practical programming experience. For all these reasons, a course on data structures cannot succeed unless there is a significant programming component.

Approach: This book describes many techniques for representing data. These techniques are presented within the context of the following principles:

1. There are costs and benefits associated with each data structure and algorithm. Practitioners need an in-depth understanding of how to evaluate the costs and benefits to be able to adapt to emerging design challenges.
2. It is rather common to reduce time requirements at the expense of increased space requirements or vice versa. Programmers regularly face trade-offs at all stages of software design and implementation, so the concept needs to be firmly anchored.
3. Programmers should be sufficiently knowledgeable about current practices to avoid reinventing the wheel. Hence, programmers need to learn the commonly used data structures, their associated algorithms, and the most frequently encountered design models found in programming.

4. Data structures respond to the requirements. Programmers need to first learn how to assess application needs and then find a data structure with matching capabilities. To do this requires competence in principles 1, 2 and 3.

I appreciate the support and encouragement of my family and friends.
I hope you get the best. Enjoy your reading, please.

Amol M. Jagtap
Ajit S. Mali

Authors

Amol M. Jagtap pursued M. Tech in Software Engineering from JNTU, Hyderabad. He is working as an Assistant Professor in the Department of Computer Science and Engineering at Rajarambapu Institute of Technology, Islampur, District Sangli. He has published more than 15 research papers in reputed international journals including IEEE etc. His main research work focuses on Machine learning and cloud computing. He has 16 years of experience in teaching and industry.

Ajit S. Mali pursued M. Tech in Computer Science and Engineering from Rajarambapu Institute of Technology, Rajaramnagar Islampur Dist. Sangli. He is currently working as an Assistant Professor in the Department of Computer Science and Engineering at RIT, Islampur. He has published more than 15 research papers in reputed international journals, conferences including IEEE, Elsevier. His main research work focuses on Cloud Computing and Internet of Things (IoT). He has 8 years of experience in teaching.

1

Fundamental Principles of Algorithm and Recursion

1.1 Algorithm, Its Pseudo-Code Representation and FlowChart

1.1.1 Algorithm Definition

An algorithm is a step-by-step procedure for solving a problem. A sequential solution to any program that is scripted in any natural language is called as an algorithm. The algorithm is the first step of the solving process. After the problem analysis, the developer writes the algorithm for this problem. It is largely used for data processing, computing and other related computer and mathematical operations.

The algorithm must satisfy the following five conditions:

1. Input: Zero, one or more quantities are externally provided.
2. Output: The algorithm produces at least one output as a result.
3. Definiteness: Each statement must be clear, specific, and unambiguous.
4. Finite: The algorithm should conclude after a limited number of steps.
5. Effectiveness: Every statement must be very fundamental so that it can be carried out by any individual using only paper and pencil.

1.1.2 Pseudo-Code Representation

The word "pseudo" means "fake," so "pseudo-code" means "fake code". Pseudo-code is an informal language used by programmers to develop algorithms. Pseudo-code describes how you would implement an algorithm without getting into syntactical details. It uses the structural conventions of a formal programming language but is proposed for human reading rather than machine reading. It is used to create an overview or outline of a

DOI: 10.1201/9781003105800-1

program. System designers write pseudo-code to make sure the programmer understands a software project's requirements and aligns the code accordingly. The pseudo-code contains no variable declaration.

The pseudo-code follows certain standard loop structures as follows:

1. FOR… ENDFOR
2. WHILE… ENDWHILE

Certain terms also exist for standard conditional clauses:

1. IF… ENDIF
2. WHILE… ENDWHILE (this is both a loop and a conditional clause by the way)
3. CASE… ENDCASE

The pseudo-code cannot be compiled within an executable program.

Example in C code:

```
if (I < 10)
{
I++;
}
```

Therefore, its respective pseudo-code is:

If I is less than 10, increment I by 1.

Benefits of pseudo-code:

1. Pseudo-code is understood by every programming language developer like Java or C#. Net developer, etc.
2. Programming languages are difficult to read for most people, but pseudo-code permits non-programmers, such as business analysts, end user or customer to review the steps to approve the projected pseudo-code that matches the coding specifications.
3. By drafting the code in human language first, the programmer protects from omitting an important step.

Example 1.1: Algorithm for Identifying the Area of a Triangle

Step 1: Start
Step 2: Take input as the base and the height of the user
Step 3: Area=(base * height)/2
Step 4: Print area of triangle
Step 5: Stop.

Example 1.2: An Algorithm Used to Determine Number Is Odd or Even

Step 1: Start
Step 2: Enter any input number
Step 3: Reminder=number mod 2
Step 4: If reminder=0, then

```
    Print "number is even"
  Else
    Print "number is odd"
  End if
```

Step 5: Stop.

1.1.3 Flowchart

The graphical representation of any program is referred to as flowcharts. A flowchart is a kind of diagram that represents an algorithmic program, a workflow or operations. A flowchart aims to provide people with a common language to understand a project or process. Developers often use it as a program-planning tool for troubleshooting a problem. Each flowchart provides the solution to a specific problem. There are a few standard graphics symbols, which are used in the flowchart as follows:

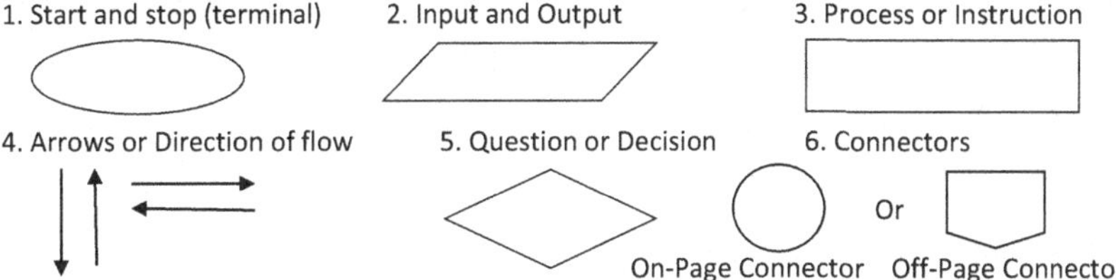

1. Start and Stop or Terminal
   The circular rectangles or oval symbol indicates where the flowchart starts and ends.
2. Input and Output
   The parallelogram symbol denotes input or output operations in the flowchart.
3. Process or Instruction
   The rectangle represents a process like computations, a mathematical operation or a variable assignment operation, etc.
4. Arrows or Direction of Flow
   Arrows or direction of flow represents the flow of the sequence of steps and direction of a process.

5. Question or Decision
   The diamond symbol is used as a representation of the true or false statement tested in a decision symbol.
6. Connector
   A flowchart is divided into two or more smaller flowcharts, consistent with project requirements. This connector is most often used when a flowchart does not fit on a single page or has to be split into sections. A connector symbol, which is a small circle called an On-Page Connector, with a number inside it, allows you to connect two flowcharts on the same page.
   A connector symbol that looks like a pocket on a shirt, called off-page connector allows you to connect to a flow diagram on another page.

Guidelines for developing flowcharts:

Here are a couple of things to keep in mind when designing and developing a flowchart:

1. The arrows should not intersect during the design of a flowchart.
2. Processes or activities take place from top to bottom or from left to right in general.
3. The on-page connectors are referenced using digits in general.
4. Off-page connectors are referenced using alphabets generally.
5. The flowchart may have only one beginning symbol and a one-stop symbol.
6. The shapes, lines and texts of a flowchart need to be consistent.

Pros and cons of flowcharts:

Advantages of flowchart:

1. Flowcharts are one of the best ways of documenting programs.
2. Flowcharts are easier to understand compared to algorithms and pseudo-codes.
3. It helps us to debug and analyze processes.
4. It helps us understand how to think or make decisions to formulate the problem.

Disadvantages of flowchart:

1. Manual tracking is needed to verify the accuracy of the paper flowchart.
2. Modification of the flowchart is sometimes time-consuming.

3. It is difficult to display numerous branches and create loops in the flowchart.
4. A simple modification of the problem logic can result in a complete redesign of the flowchart.
5. It is very difficult to draw a flowchart for huge and complicated programs.

Example: Draw a flowchart to find area of triangle and odd or even number:

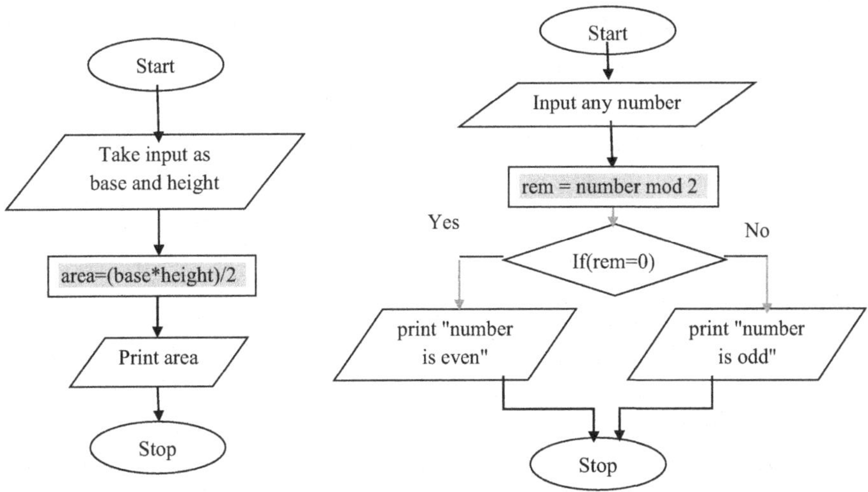

1.2 Abstract Data Type

Abstract data type (ADT) is a type or class of objects whose behavior is determined by a set of values and a set of functions. The definition of ADT only refers to the transactions that need to be performed, but not the way these transactions will be implemented. It does not specify how the data will be organized in memory and what algorithms will be used for the implementation of those operations. It is called "abstract" because it provides an independent perspective from the implementation. The process of supplying only the essentials and concealing details is known as abstraction.

The programmer who has used the data type may not know that how data type is implemented, for example, we have been using different built-in data types such as int, long, float, double, char only with the knowledge that which values we can store in that particular data types and operations that can be performed on them without any idea of how these data types are implemented. Therefore, a programmer just needs to know what a kind of data can do, but not how it's going to do it. ADT may be considered a black box that hides the internal structure and design of the data type. We are now going to define three ADTs: List ADT, Stack ADT, Queue ADT.

1.2.1 List ADT

A list contains items of the same type organized in sequence, and the following operations can be carried out on the list.

Get () – Returns an element of the list to a specific position.

Insert () - Insert an item in any position on the list.

Remove () – Remove the first instance of an item in a non-empty list.

RemoveAt () – Removes the element at a specified location from a non-empty list.

Replace () – Replace an item at any position with a different item.

Size () – Returns the number of items within the list.

IsEmpty () – Returns true if the list is empty, else returns false.

IsFull () - Returns true if the list is full, otherwise it returns false.

1.2.2 Stack ADT

A stack contains elements of the same type arranged in sequential order and push and pop operations occur at one end at the top of the stack. The following operations may be carried out on the stack:

Push () – Insert an item on one end of the stack called the top.

Pop () – Removes and returns the item to the top of the stack, if it is not empty.

Top () – Returns the element to the top of the stack without deleting it, if the stack is not empty.

Size () – Returns the number of items within the stack.

IsEmpty () – Returns true if it is empty, otherwise returns false.

IsFull () – Returns true if the stack is full or returns false.

1.2.3 Queue ADT

A queue contains similar items organized in a sequential order. The operations take place at both ends, which is at the front and rear, the insertion is done at the rear and the deletion is done at the front. The following operations may be carried out in the queue:

Enqueue () – Insert an element at the end of the queue that is the rear end.

Dequeue () – Delete the first item from the front end of the queue, if the queue is not empty.

Size () – Returns the number of items within the queue.

IsEmpty () - Returns true if the queue is empty, otherwise return false.

IsFull () – Returns true if the queue is full or returns false.

From these descriptions of ADTs, we can see that the descriptions do not specify how these ADTs will be represented and how the operations will be carried out. There can be alternative ways to implement ADTs using an array or linked list, for example, the list ADT can be implemented using arrays as statically, or singly linked list or doubly linked list as dynamically. Likewise, stack ADT and Queue ADT or other ADTs can be implemented using related arrays or linked lists.

1.3 Data Structure

1.3.1 Definition

The data structure is a collection of data elements organized in a specific manner and functions is defined to store, retrieve, delete and search for individual data elements.

All data types, including built-in such as int and float; derived such as array and pointer; or user defined such as structure and union, are nothing but data structures. A data structure is a specialized format for organizing and storing data, with features for recovering, removing and search for data items. General data structure includes array, list, stack, tree, graph, etc. Each data structure is designed to organize data with particular intention so that it can be accessed and used appropriately.

The data structure study covers the following topics:

1. Amount of storage needed.
2. Amount of time required to process the data.
3. Data representation of the primary memory.
4. Operations carried out with such data.

1.3.2 Types of Data Structures or Their Classification

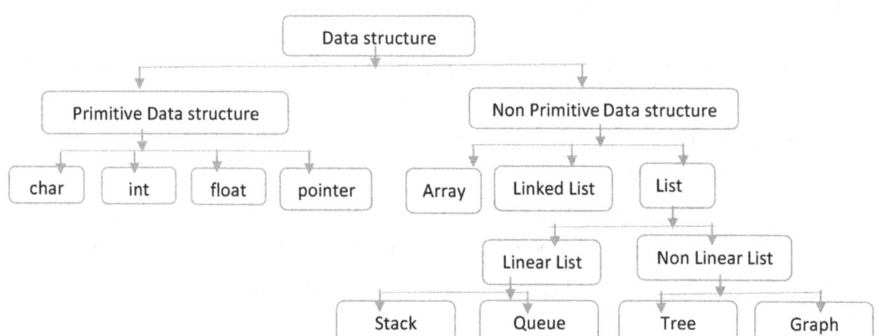

1.3.3 Difference in Abstract Data Types and Data Structures

The ADT is the interface. It is defined by what operations it will support. It does not specify how the data structure or algorithmic technique will be used to perform these operations.

For example, many different ideas can be used to implement a queue: it could be a circular queue, linear queue, doubly ended queue and priority queue. Queue implemented using arrays or linked lists. As a result, the data structure is an implementation of the ADT.

ADT is a logical picture of data and data manipulation operations. The data structure is the genuine representation of the data during implementation and the algorithms for manipulating the data elements. ADT is the logical level, and the data structure is an implementation level.

ADT is a representation of a data type in terms of possible values and not how the values are stored in memory, operations permissible without concerning its implementation details.

So stack when implemented using a programming language like C, would be a data structure, but describing it in terms of things like its behavior such as last in, first out (LIFO), a possible set of values, operations permissible such as push, pop, etc. would be considered as an ADT for stack.

1.4 Algorithm Efficiency or Performance Analysis of an Algorithm

The effectiveness of the algorithm or the analysis of the performance of an algorithm mainly concerns two things: the complexity of space and the complexity of time.

1.4.1 Space Complexity

Space complexity is defined as the total amount of primary memory a program needs to run to its completion.

1.4.1.1 Components of Program Space

Program space covers instruction space, data space and stack space depicted as follows:

Program space=Instruction space+data space+stack space.

1. **Instruction space is dependent on several factors:**
   i. A compiler that generates the machine or binary code
   ii. Target computer

2. **Data space:**
   i. Heavily dependent on computer's architecture and compiler.
   ii. Magnitude of data that program works.

   In programming languages, there are different kinds of data and different number of bytes to store particular data are required. Such that the character data type requires 1 byte of memory, integer requires 4 bytes, float requires 4 bytes, double requires 8 bytes, etc. This varies from one compiler to another. Choosing a "smaller" data type will affect the overall use of space in the program. Choose the appropriate type when working with arrays.

3. **Stack space:**

   Each time a function is called, the following data are recorded in the stack.
   i. The return address.
   ii. Value of all local variables and value of formal parameters.
   iii. Bindings of all references and constant reference parameters.

   A few points to make efficient use of space:
   i. Always select the smallest data type that you need.
   ii. Go through the compiler in detail.
   iii. Discover the effects of various compilation settings.
   iv. Choose a non-recursive algorithm, if applicable.

   Space (p) requirements for any p algorithm can therefore be written as,

S (p)=c+Sp (instance characteristics)

C=fixed part: this includes instruction space, space for single and fixed-size variables, constants, etc.

Sp=variable part: space needed for referenced variables, which depends on instance characteristics and recursion state space.

Examples:

1. Non-recursive algorithm sum(a, n)

```
{
S = 0.0;
for   i =1 to n do
S = S + a[i] ;
return S;
}
```

Here the space complexity of non-recursive sum algorithm is as follows:

```
Ssum(n) >= (n+3)
```

where Ssum (n) means the space complexity of non-recursive algorithm sum having size of array elements n.

And (n+3) means n for array element a[], one each for n, i and s.

2. Recursive algorithm Rsum(a, n)

```
{
if (n<=0) then return 0.0;
else return Rsum (a, n-1) + a[n];
}
```

Space required for the recursive Rsum algorithm is:

Each call to Rsum requires at least three words such as each for n, return address and a pointer to array a []. Recursion call occurs (n+1) times so the recursive state space needed is ≥ 3 (n+1).

That means the space complexity of SRsum (n) ≥ 3 (n+1) where SRsum (n) means the space complexity of recursive algorithm Rsum having size of array elements n.

1.4.2 Time Complexity

The amount of time it takes for an algorithm to run is called as time complexity.

Absolute time complexity is defined as follows:

Absolute time complexity = **number of instructions**

$\times$ **time taken to execute one instruction**

However, the time taken to execute one instruction, this quantity varies from system to system or computer to computer. Therefore, we only consider the number of instructions in an algorithm to be a time complexity.

For most algorithms, running time depends on the size of the input. Run time is expressed in T (n) for a function T on input size n. Suppose there is if….. else statement that may execute or not, then in that situation variable running time.

In general, algorithm time complexity is calculated in the following ways:

1. **Worst case:** The maximum time required to execute the program. The time complexity of the worst case is represented by the big Oh notation, O.
2. **Average case:** The average time required to execute the program. The time complexity of the average case is represented by the Theta notation, θ.
3. **Best case:** The minimum time required to execute the program. The time complexity of the best case is represented by Omega notation, Ω.

The efficiency or analysis of an algorithm involves two phases:

1. **Prior estimate:**

   Prior estimates concern theoretical or mathematical calculations. Theoretical or mathematical calculations use the asymptotic notation for representation.
2. **Posterior estimate or testing:**

   The designed algorithm is implemented on a machine. It deals with the writing of a program and thus depends on the operating system, compiler, etc.

Performance analysis: The primary goal is to determine the time and space required to run the program to its completion in terms of system time and system memory.

Prior analysis: Deals with computations of count of statement in a given algorithm. The count indicates how often each instruction is executed.

1. Example 1

```
Algorithm add (a, n)
{
```

```
3. S=0;
4. for (i=1 to n) do
5. S = S + a[i];
6. return s;
7. }
```

Line Number	Step per Execution	Frequency	Total Steps
1.	0	0	0
2.	0	0	0
3.	1	1	1
4.	1	n+1	n+1
5.	1	N	N
6.	1	1	1
7.	0	0	0
			2n+3 step count

2. Example 2:

```
1. Algorithm add(a,b,c,m,n)
2. {
3. for (i=1 to m) do
4. for (j=1 to n) do
5. C[i, j]=a[i, j]+b[i, j]
6. }
```

Line Number	Total Steps
1.	0
2.	0
3.	m+1
4.	m(n+1)
5.	Mn
6.	0
	2mn+2m+1 step count

In examples 1 and 2, the number of steps carried out is shown. There are three types of **step count:**

a. **Best case:** Minimum number of steps that can be executed for given inputs.

b. **Worst case:** Maximum number of steps that can be executed for given inputs.

c. **Average case:** Average number of steps that can be executed for given inputs.

1.5 Recursion and Design of Recursive Algorithms with Appropriate Examples

The recursive function is a function that is invoked by itself. Similarly, an algorithm is said recursive if the same algorithm is called within the body.

Recursive algorithms can be divided into the following types:

1. **Direct recursive algorithm:** An algorithm is said to direct recursive algorithm if the same algorithm is called within the body.

   Example:

```
#include<stdio.h>
int main()  // main function is called as direct
recursive function.
{
        printf("Hello\n");
        main();          // call to the main function
        return 0;
}
```

   Output:

   Print Hello infinite times.

   In the above program, the main () function is invoked by itself is called as direct recursive function.

2. **Indirect recursive algorithm:** Algorithm A is considered an indirect recursive algorithm if it calls another algorithm B that in turn calls A.

   Example:

```
#include<stdio.h>
void demo();
int main()   //main function is called as indirect
recursive function.
{
printf("Hello\n");
demo();
return 0;
}
void demo()
{
main();     // call to the main function inside demo
function and demo function is called by  // main
function
}
```

   Output:

Print Hello infinite times.

In the above program, the demo () function is invoked by the main () function and the main function is again invoked inside the body of the demo () function is called as indirect recursive function.

1.5.1 Example 1: Implementing a Factorial Program Using Non-recursive Functions

```
#include <stdio.h>
long factorial(int);
int main()
{
int number;
long fact = 1;
printf("Enter a number to calculate its factorial\n");
scanf("%d", &number);
printf("%d! = %ld\n", number, factorial(number));
return 0;
}
long factorial(int n)
{
int c;
long fact = 1;
for (c = 1; c <= n; c++)
{
fact = fact * c;
}
return fact;
}
```

Output:

Enter a number to calculate its factorial
4
4!=24

1.5.2 Example 2: Implementing a Factorial Program Using Recursive Functions

```
#include<stdio.h>
long factorial(int);
int main()
{
int n;
long f;
printf("Enter an integer to find its factorial\n");
scanf("%d", &n);
if (n < 0)
printf("Factorial of negative integers isn't defined.\n");
```

```
else
{
f = factorial(n);
printf("%d! = %ld\n", n, f);
}
return 0;
}
long factorial(int n)
{
if (n == 0)
return 1;
else
return(n * factorial(n-1));
}
```

Output:

Enter an integer to find its factorial
4
4!=24

1.5.3 Example 3: Implementing a Program That Displays the Sum of Digits in a Number Using Non-recursive Functions

```
#include<stdio.h>
int sum_of_digit(int n)
{
int r,sum =0;
r = n % 10;    // separate first digit of a number
sum = sum + r;    // add the separated digit in the
intermediate variable
n = n/10;    // delete first digit of a number
}
int main()
{
int num = 123;
int result = sum_of_digit(num);
printf("Sum of digits in %d is %d\n", num, result);
return 0;
}
```

Output:

The sum of digits in 123 is 6.

1.5.4 Example 4: Implementing a Program That Displays the Sum of Digits in a Number Using Recursive Functions

```
#include<stdio.h>
int sum_of_digit(int n)
```

```
{
if (n == 0)
return 0;
return (n % 10 + sum_of_digit(n / 10));
}
int main()
{
int num = 123;
int result = sum_of_digit(num);
printf("Sum of digits in %d is %d\n", num, result);
return 0;
}
```

Output:

Sum of digits in 123 is 6

1.5.5 Example 5: Implementation of the Sum of n First Natural Numbers by Means of Non-recursive Functions

```
#include <stdio.h>
int addNumbers(int n);
int main()
{
int num;
printf("Enter a positive integer: ");
scanf("%d", &num);
printf("Sum = %d",addNumbers(num));
return 0;
}
int addNumbers (int n)
{
int i, sum=0;
for(i=1; i<=n; i++)
{
sum = sum + i;
}
return sum;
}
```

Output:

Enter a positive integer: 5
Sum=15

1.5.6 Example 6: Implementation of the Sum of n First Natural Numbers by Means of Recursive Functions

```
#include <stdio.h>
int addNumbers(int n);
```

```
int main()
{
int num;
printf("Enter a positive integer: ");
scanf("%d", &num);
printf("Sum = %d",addNumbers(num));
return 0;
}
int addNumbers (int n)
{
if(n != 0)
return n + addNumbers(n-1);
else
return n;
}
```

Output:

Enter a positive integer: 5
Sum=15

1.5.7 Example 7: Implementation of the Prime Number Program Using Non-recursive Functions

```
#include<stdio.h>
int isPrime(int,int);
int main()
{
int num, prime;
printf("Enter a positive number: ");
scanf("%d",&num);
prime = isPrime(num, num/2);
if(prime==1)
        printf("%d is a prime number",num);
else
      printf("%d is not a prime number",num);
return 0;
}
int isPrime(int num, int n)
{
    int i;
       for(i=n; i>=2 ; i--)
    {
       if(num % i==0)
       {
       return 0;  // if number is not prime
       }
}
       return 1;   // if number is prime
}
```

Output:

Enter a positive number: 7
7 is a prime number

1.5.8 Example 8: Implementation of the Prime Number Program Using Recursive Functions

```
#include<stdio.h>
int isPrime(int,int);
int main()
{
int num, prime;
printf("Enter a positive number: ");
scanf("%d",&num);
prime = isPrime(num, num/2);
if(prime==1)
        printf("%d is a prime number",num);
else
      printf("%d is not a prime number",num);

return 0;
}

int isPrime(int num, int i)
{
if( i==1 )
{
        return 1;
}else{
       if(num % i ==0)
         return 0;
       else
         isPrime(num, i-1);
        }
}
```

Output:

Enter a positive number: 21
21 is not a prime number

1.5.9 Example 9: Implementation of the Reverse Number Program Using Non-recursive Functions

```
#include<stdio.h>
int reverse_function(int num);
int main()
{
   int num, reverse_number;
   printf("\nEnter any number:");
   scanf("%d",&num);
```

```
   reverse_number=reverse_function(num);
   printf("\nAfter reverse, the number is
:%d",reverse_number);
   return 0;
}
int rev=0, rem;
int reverse_function(int num)
{
   while(num!=0)
   {
      rem=num % 10;    // separate first digit of a number
      rev=(rev * 10 )+ rem;    //multiply each intermediate
number (rev) by 10 and add separated //digit into it gives you
reverse number
      num = num / 10;  // delete first digit of a number
   }
      return rev;
}
```

Output:

Enter any number: 546
After reverse, the number is: 645

1.5.10 Example 10: Implementation of the Reverse Number Program Using Recursive Functions

```
#include<stdio.h>
int reverse_function(int num);
int main()
{
   int num, reverse_number;
   printf("\nEnter any number:");
   scanf("%d",&num);
   reverse_number=reverse_function(num);
   printf("\nAfter reverse, the number is
:%d",reverse_number);
   return 0;
}
int sum=0, rem;
int reverse_function(int num)
{
   if(num)
   {
      rem=num%10;
      sum=sum*10+rem;
      reverse_function(num/10);
   }
   else
      return sum;
}
```

Output:

Enter any number: 123
After reverse, the number is: 321

1.6 Interview Questions

1. What is an algorithm and flowchart explain with suitable example?
2. Explain pseudo-code representation in detail.
3. Write an algorithm to read three different numbers from the user and then find the largest between that three numbers.
4. Draw the flowchart to check whether a number is a prime number.
5. Write an algorithm and flowchart for finding the sum of all odd numbers from 100 to 500.
6. Write an algorithm and flowchart to calculate even numbers between 0 and 500.
7. Write an algorithm and flowchart to find the middle number by giving three numbers.
8. What is an ADT in data structure?
9. What is the difference between an ADT and a data structure?
10. Explain data structure with suitable examples?
11. What are the applications of data structures?
12. What are the different types of data structures?
13. What are the different advantages and disadvantages of flowchart?
14. What is the performance analysis of an algorithm?
15. What are the two main measures for the efficiency of an algorithm?
16. Explain the different symbols used in the flowchart.

1.7 Multiple Choice Questions

1. Which of the following issues cannot be resolved by recursion?
   A. Factorial of a number
   B. Finding of a prime number

C. **Problems without base case**

D. Finding of sum of digits in any number

Answer: (C)
Explanation:
Problems having a base case should be solved using recursion. However, problems without base case lead to infinite recursion call and cannot be resolved using recursion.

2. The recursion process is identical with which of the following construct?

A. Switch case

B. If else

C. **Loop**

D. if else if ladder

Answer: (C)
Explanation:
The recursion process is like a loop and all recursive calls are kept in the memory area of the stack.

3. What will be the output of the following code?

```
#include<stdio.h>
void recursion(int n)
{
    if(n == 0)
    return;
    printf("%d",n);
    recursion(n-2);
}
int main()
{
    recursion(6);
    return 0;
}
```

A. **642**

B. 654321

C. 662

D. 6420

Answer: (A)
Explanation:
Recursion function's first call print value 6, second call print value 4, third call print value 2 and the fourth call will execute return statement when the value becomes 0, so the output of the program is 642.

4. Which of the following data structure used to implement recursion in the main memory?
   A. Stack
   B. Array
   C. Linked list
   D. Tree

   Answer: (A)
   Explanation:
   Stack is a data structure that is used to implement recursion into the main memory.

5. What will be the output of the following C code?

```
#include<stdio.h>
int main()
{
    printf("Hello students");
    main();
    return 0;
}
```

   A. Hello students is printed once
   B. Hello students is printed infinite times
   C. Hello students is printed till the stack overflows occur
   D. None of the above

   Answer: (C)
   Explanation:
   In the above code, the main () function itself, calling main () is called as recursion. In the above code, there is no termination condition to exit from the program. Hence, "Hello students" will be printed an infinite number of times. However, practically, when the stack memory partition is overflowing, the printing of "Hello students" will be stopped.

6. A sequential solution of any program that is written in any natural language is called as:
   A. Flowchart
   B. Algorithm
   C. Method
   D. A and B both

   Answer: (B)

Explanation:

A sequential solution of any program that wrote in any natural language is called as an algorithm.

7. Which of the following code describes how you would implement an algorithm without getting into syntactical details.
   A. Program code
   B. **Pseudo-code**
   C. Machine-level code
   D. Binary code

   Answer: (B)

   Explanation:

   Pseudo-code describes how you would implement an algorithm without getting into syntax details.

8. Which of the following is true for ADT that is ADT?
   I. ADT is a type or class for objects whose behavior is defined by a set of values and a set of functions.
   II. ADT is only stating that what operations are to be carried out.
   III. ADT does not mention how data will be organized in memory.
      A. I and II are correct
      B. Only I is correct
      C. II and III are correct
      D. I, II and III are correct

   Answer: (D)

   Explanation:

   ADT is a type or class for objects whose behavior is defined by a set of values and a set of functions. ADT only mentions what operations are to be performed, but not how these operations will be implemented.

9. Algorithm efficiency or performance analysis of an algorithm is concerned with which of the following things.
   A. Time complexity
   B. Space complexity
   C. Both A and B
   D. None of the above

   Answer: (C)

   Explanation:

The effectiveness of the algorithm or the analysis of the performance of an algorithm is measured in terms of time and space complexity.

10. Elements of program space within space complexity include which of the following.
    A. Instruction space
    B. Data space
    C. Stack space
    D. **All the above**

    Answer: (D)
    Explanation:
    Program space=Instruction space+data space+stack space.

11. Which of the following statements is false or true?
    1. Pseudo-code is an informal language used by programmers to develop algorithms.
    2. The pseudo-code contains no variable declaration.

    A. Statement 1 is false
    B. Statement 2 is false
    C. Statements 1 and 2 both are false
    D. **Statements 1 and 2 both are true**

    Answer: (D)

12. Which of the following statements is false or true?
    1. A flowchart is a kind of diagram which represents an algorithmic program, a workflow or operations.
    2. A sequential solution of any program that is scripted in any natural language is called as an algorithm.
    A. Statement 1 is false
    B. Statement 2 is false
    C. Statements 1 and 2 both are false
    D. **Statements 1 and 2 both are true**

    Answer: (D)

13. Which of the following statements is false or true?
    1. Space complexity is defined as the total amount of secondary memory a program needs to run to its completion.

2. The amount of time required for a program to compile is called as its time complexity.
   A. Statement 1 is false
   B. Statement 2 is false
   C. Statements 1 and 2 both are false
   D. Statements 1 and 2 both are true

   Answer: (C)

14. Which of the following statements is false or true?
    1. Minimum number of steps that can be executed for given inputs is called as best case.
    2. Maximum number of steps that can be executed for given inputs is called as worst case.
       A. Statement 1 is false
       B. Statement 2 is false
       C. Statements 1 and 2 both are false
       D. **Statements 1 and 2 both are true**

    Answer: (D)

15. Which of the following statements is false or true?
    1. The recursive function is a function that is invoked by itself.
    2. Algorithm A is considered direct recursive algorithm if it calls another algorithm B that in turn calls A.
       A. Statement 1 is false
       B. Statement 2 is false
       C. Statements 1 and 2 both are false
       D. Statements 1 and 2 both are true

    Answer: (B)

2

Sequential Representation of Linear Data Structures

2.1 Distinction between Linear Data Structure and Nonlinear Data Structure

Linear Data Structure

Data structure where the data elements are organized sequentially or linearly where the elements are attached to its preceding and subsequent adjacent element is called a linear data structure. In the linear data structure, single level is involved. Therefore, we can traverse all the elements in single run only. Linear data structures are easy to implement as computer memory is organized linearly. Examples include arrays, stack, queue, linked list, etc.

Nonlinear Data Structure

Data structures in which data elements are not organized sequentially or linearly are referred to as nonlinear data structures. In a nonlinear data structure, single level is not involved. Therefore, it is not possible to go through all the elements at once. Nonlinear data structures are not easy to implement with respect to the linear data structure. It makes effective use of computer memory in relation to a linear data structure. Examples include trees and graphs.

2.2 Operations on Stack

Stack is a non-primitive linear data structure. Stack is an ordered list, but access, insertions and deletions of elements are restricted by following certain rules.

DOI: 10.1201/9781003105800-2

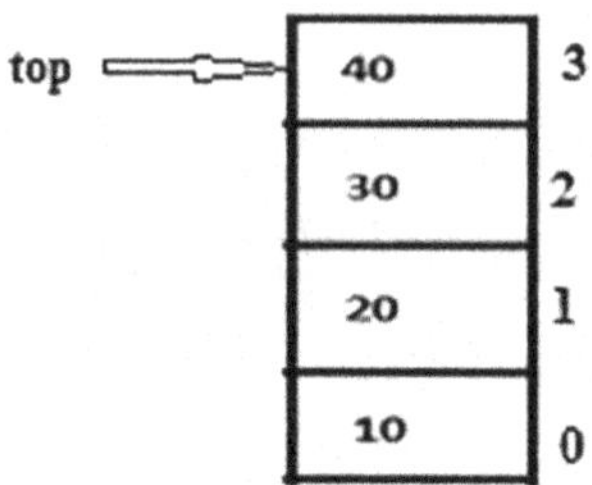

FIGURE 2.1
Stack.

Example: In the stack, pop the element which is last to be placed on the stack and push the element at the top of the stack, such an order is called last in first out (LIFO).

Concept of stack:

A stack is an ordered list in which all insertions and deletions are performed at one end, referred to as top (Figure 2.1).

2.2.1 Primitive Operations on Stack

1. To create a stack: Create a stack in memory. Creation of stack can be done either using arrays or using linked lists.
2. To insert an element onto the stack that is pushing operation
3. To delete an element from stack that is the pop operation
4. To verify which element is at the top of the stack
5. To verify whether a stack is empty or full.

Representing the stack using arrays:

Declaration 1 (Figure 2.2):

```
#define size 100
int  stack[size],top = -1;
```

Declaration 2 (Figure 2.3):

```
#define size 10
struct stack
{
```

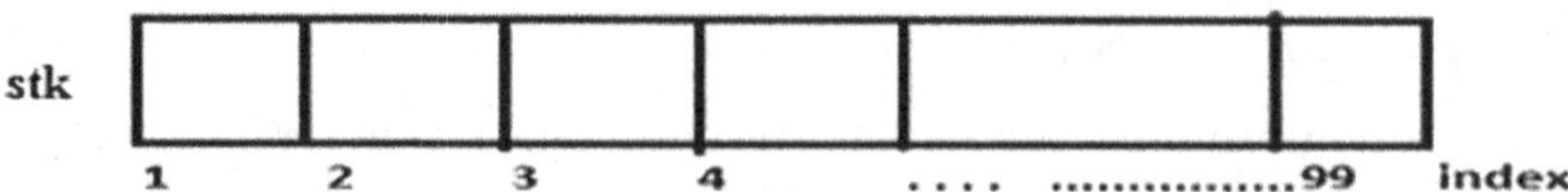

FIGURE 2.2
Stack implementation using one-dimensional array data type.

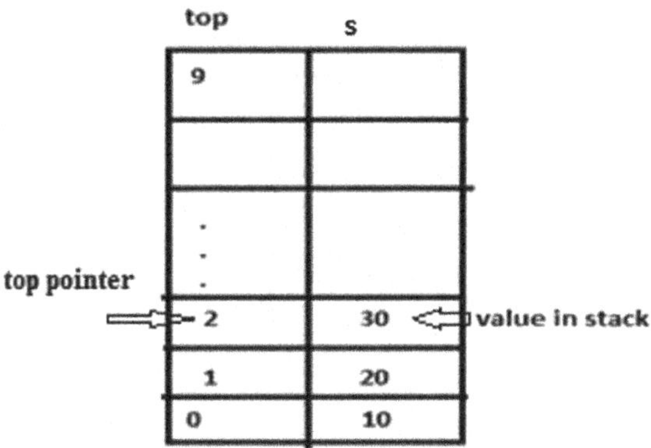

FIGURE 2.3
Stack implementation using structure data type.

```
        int s[size];
        int top;
} st ;
```

We will write push and pop function to implement the stack, and before pushing, we will check whether or not stack is full. Before popping an element from stack, we will check whether or not the stack is empty.

Empty stack operation (Figure 2.4):

Top initially set as –1.

```
As  st.top = -1 ;  // when your stack is empty
int  stack_empty()
{
        if(st.top == -1)
        return  1:
        else
        return 0;
}
```

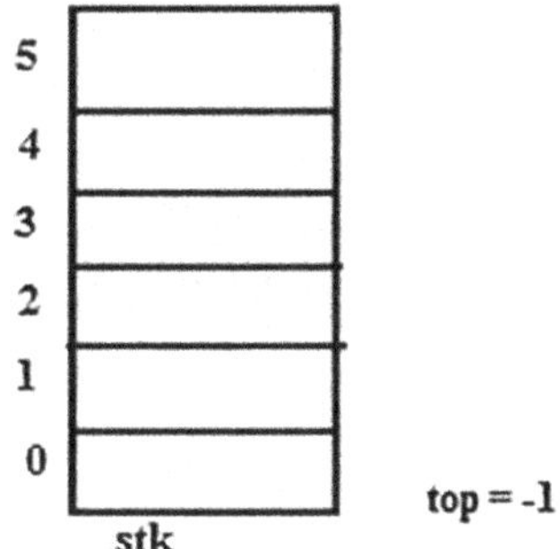

FIGURE 2.4
Stack empty condition.

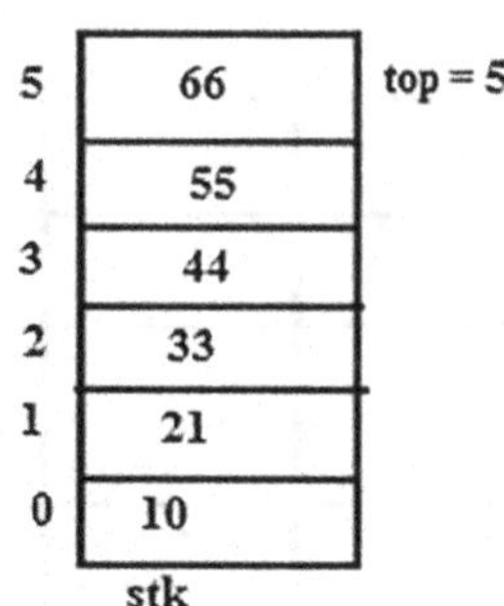

FIGURE 2.5
Stack full condition.

Stack full operation (Figure 2.5):

```
        int stack_full()
{
        if(st.top == (size-1))
        return 1;
        else
        return 0;
}
```

If the stack is full it returns 1 otherwise it returns 0.

2.2.2 Algorithm for Push and Pop Function

An algorithm for the Push operation: push (int item) inserts the element to the top of the stack [max size].

Step 1: Start

Step 2: Initialize
    Set top= -1

Step 3: Repeat steps 3 to 5 until top < max size-1

Step 4: Read item.

Step 5: Set top = top +1

Step 6: Set stack [top] = item

Step 7: Otherwise Print "stack is overflow"

Step 8: Stop

```
void push(int item)
{
          st.top++;
          st.s[st.top]=item;
}
```

An algorithm for the Pop operation: the pop ( ) removes the element from the top of the stack (Figure 2.6).

Step 1: Start

Step 2: Repeat step 3 to 5 until top >= 0

Step 3: Set item = Stack[top]

Step 4: Set top = top -1

Step 5:Print, No deleted is, Item.

Step 6: Otherwise Print stack under flows.

Step 7: Stop

```
int pop()
{
```

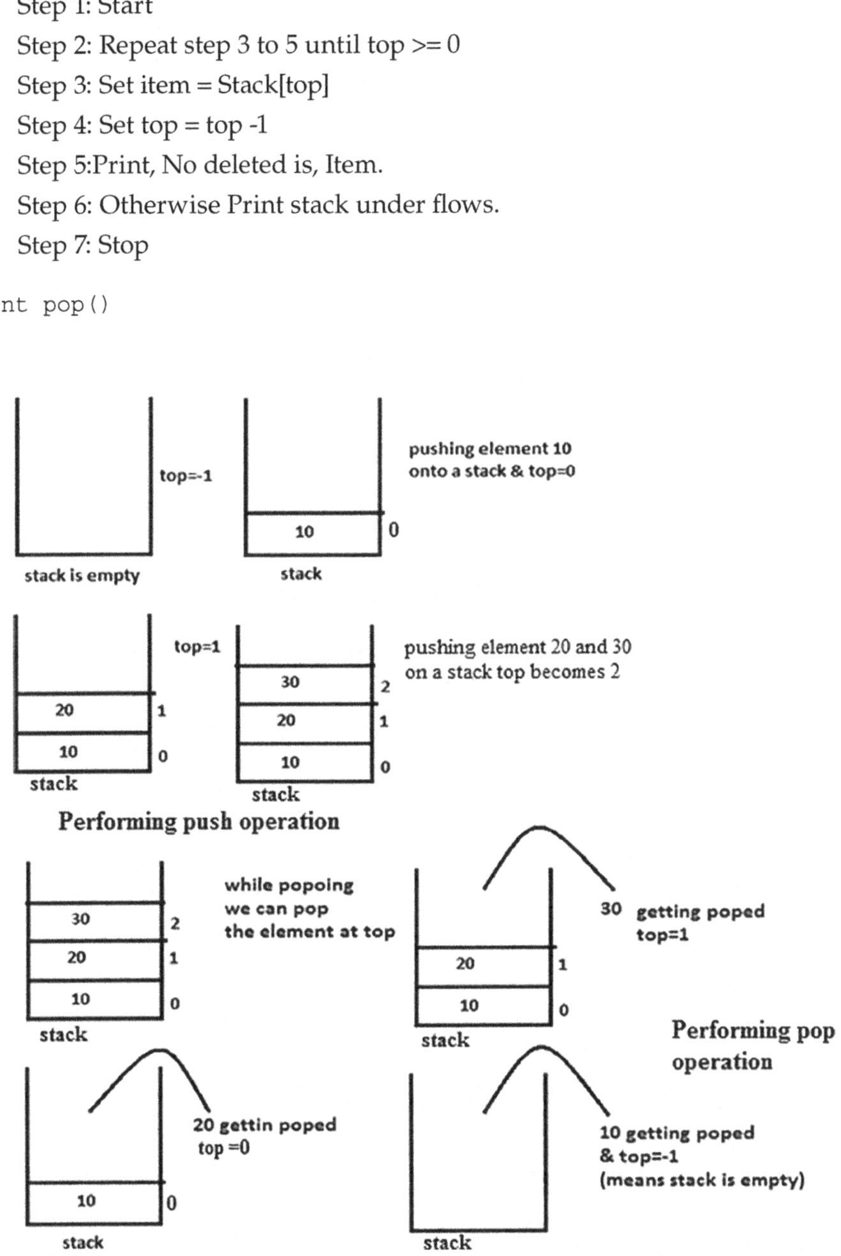

FIGURE 2.6
Performing push and pop operation on a stack.

```
        int item;
        item=st.s[st.top];
        st.top--;
        return(item);
}
```

2.2.3 A C Program for Stack Implementation Using the Array

```
#include<stdio.h>
#include<stdlib.h>
#define max 5
int stack[max];
int top;
//function declaration
void push(int);
int pop();
void display();
//main function
int main()
{
 int choice;
 int data;
 char ch;
 top=-1;
 printf("Stack Implementation using array: \n");
 do
 {
  printf("1.push\t2.pop\t3.display\t4.exit");
  printf("\nEnter your choice:");
  scanf("%d", &choice);
  switch(choice)
  {
   case 1:
 printf("Enter data:");
       scanf("%d", &data);
       push(data);
       break;
   case 2:
       data=pop();
       printf("popped element=%d\n",data);
       break;
   case 3:
       display();
       break;
   case 4:
       exit(0);
   default:
       printf("\n wrong choice");
  }
fflush(stdin);  // clears the standard input buffer
```

```
printf("Do you want to continue[y/n] :");
scanf("%c",&ch);
}while(ch=='y');
return 0;
}
//insertion of an element into the stack
void push(int a)
{
 //check stack is full
if(top==max-1)
{
printf("\n stack is full");
return;
}
 //space in the stack
else
{
top++;
stack[top]=a;
}
}
//delete an element from the stack
int pop()
{
 int item;
 //check stack is empty
if(top==-1)
{
printf("\n stack is empty");
}
//the stack is created
else
{
item=stack[top];
top--;
}
return(item);
}
//display the stack elements
void display()
{
int i;
//check stack is empty
if(top==-1)
{
printf("\n stack is empty");
}
//element is present
else
{
```

```
printf("Elements in the stack are : ");
for(i=top; i>=0; i--)
{
printf("\t%d", stack[i]);
}
printf("\n");
}
}
```

Output:

```
Stack Implementation using array:
1.push  2.pop   3.display       4.exit
Enter your choice:1
Enter data:21
Do you want to continue[y/n] :y
1.push  2.pop   3.display       4.exit
Enter your choice:1
Enter data:31
Do you want to continue[y/n] :y
1.push  2.pop   3.display       4.exit
Enter your choice:1
Enter data:41
Do you want to continue[y/n] :y
1.push  2.pop   3.display       4.exit
Enter your choice:3
Elements in the stack are :     41      31      21
Do you want to continue[y/n] :y
1.push  2.pop   3.display       4.exit
Enter your choice:2
popped element=41
Do you want to continue[y/n] :y
1.push  2.pop   3.display       4.exit
Enter your choice:3
Elements in the stack are :     31      21
Do you want to continue[y/n] :y
1.push  2.pop   3.display       4.exit
Enter your choice:4
```

2.3 Applications of Stack

2.3.1 Expression Evaluation

In expression evaluation, stack is used.

2.3.2 Expression Conversion

a. Infix to postfix conversion:

(a + b) * (c - d) given infix notation is converted into postfix form a b + c d - * using operator and operand stack.

b. Infix to prefix conversion:

(a + b) * (c - d) given infix notation is converted into prefix form * + a b - c d using operator and operand stack.

c. Postfix to infix conversion:

A B - C *

Solution:

The first given formula is in the postfix form, therefore **scan the formula from left to right.**

Algorithm

1. Check the character is operand or operator.
2. If it is an operand, then push it on the operand stack.
3. If it is an operator, then place it in the operator stack; then check whether it is a unary or binary operator.
4. If it is a unary operator, then pop current one operand from the operand stack and do the operation on that operand and result which is operand itself is stored on the operand stack.
5. If it is a binary operator, then pop current two operands from the operand stack and do the operation on that operands and result which is operand itself is stored on the operand stack.
6. Do the same procedure till the character sting get over.

Operator Stack	Operand Stack
	A
	B
-	
	(A - B)
	C
*	
	((A - B) * C)
	This is completely parenthesized infix form.

d. Prefix to infix conversion:

Convert the following expression from prefix to infix:

- + a b * c d

Solution:

The first given formula is in the prefix form, therefore scan the string or formula from right to left.

Algorithm

1. Check the character is operand or operator.
2. If it is an operand, then push it on the operand stack.
3. If it is an operator, then place it in the operator stack; then check whether it is a unary or binary operator.
4. If it is a unary operator, then pop current one operand from the operand stack and do the operation on that operand and result, which is operand itself is stored on the operand stack.
5. If it is a binary operator, then pop current two operands from the operand stack and do the operation on that operands and result, which is operand itself is stored on the operand stack.
6. Do the same procedure till the character sting get over.

Operator Stack	Operand Stack
	D
	C
*	
	(c * d)
	B
	A
+	
	(a + b)
-	
	((a + b) - (c * d)) **This is completely parenthesized infix form.**

2.3.3 Backtracking

Backtracking is used in algorithms where there are steps along a certain path from a point of departure to a certain target.

1. Find your path through the maze.
2. Find a route from one point in a graph or roadmap to another point.
3. Play a game in which there are moves to be made in checkers or chess.

In all of these cases, there are alternatives to be made among several options. We need some way to remember these decision points in case we want to come back and try the alternative.

Consider the maze. At a point where an alternative is made, we may find out that the choice leads to a dead-end. We want to retrace back to that decision point and then try the next alternative.

In the above example, stack, data structures can be used as part of the solution of the backtracking.

2.3.4 Function Call and Function Return Process

When a method or function is called which, is a function call then,

1. An activation record is created; its size depends on the number and size of the local variables and local parameters.
2. The base pointer value is saved in the particular location reserved for it.
3. The program counter value is stored in the return address location.
4. The base pointer is now reset to the new base pointer.
5. Copies the calling parameters in the region of the parameters.
6. Initialize the local variables within the area of the local variable.

For maintaining function call data stack data structure is used.

2.3.5 Recursive Functions

Recursion uses stack memory to store intermediate results of all local variables for that particular function call.

2.3.6 Depth First Search (DFS)

Stack data structure is used in DFS.

2.3.7 Undo Mechanism in Text Editors

It is implemented using stack data structure.

2.3.8 Reverse the String

In reverse string application we first push all characters in a string on the stack and then pop the characters from the stack gives you reverse string.

2.3.9 Towers of Hanoi Problems

It is solved by using stack.

2.3.10 Parsing or Syntax Analysis

Parsing or syntax analysis is the process of analyzing whether a string of formal language statements conforms to the grammar or not. For this purpose, we have to create a parse tree or derivation tree, and in this parsing or syntax analysis, we have to use stack data structure.

2.4 Implementing Stack Applications

2.4.1 Implementation of Post-Fix to Infix Conversion Using Stack

Program:

```
#include<stdio.h>
#include<string.h>
#include<stdlib.h>
#define MAX 20
char str[MAX],stack[MAX];
int top=-1;
void push(char c)
{
   stack[++top]=c;
}
char pop()
{
   return stack[top--];
}
void post_in()
{
   int n,i,j=0;
   char a,b,op,x[20];
   printf("Enter the postfix expression\n");
   fflush(stdin);
   gets(str);
   strrev(str);
   n=strlen(str);
   for(i=0;i<MAX;i++)
      stack[i]='\0';
   printf("Infix expression is: \t");
   for(i=0;i<n;i++)
   {
      if(str[i]=='+'|| str[i]=='-'|| str[i]=='/'|| str[i]=='*')
      {
      push(str[i]);
      }
      else
      {
```

```
      x[j]=str[i];
      j++;
      x[j]=pop();
      j++;
      }
   }
   x[j]=str[top--];
    strrev(x);
   printf("%s\n",x);
}
int main()
{
int ch;
while(1)
{
    printf("Enter choice 1 postfix to infix 2.exit\n");
    scanf("%d",&ch);
    switch(ch)
    {
     case 1:
     post_in();
     break;
     case 2:
     exit(0);
     default: printf("Wrong chioce\n");
   }
   printf("Enter 1 to continue, 0 to exit\n");
   scanf("%d",&ch);
   if(ch==0)
    {
    break;
     }
 }
return 0;
}
```

Output:

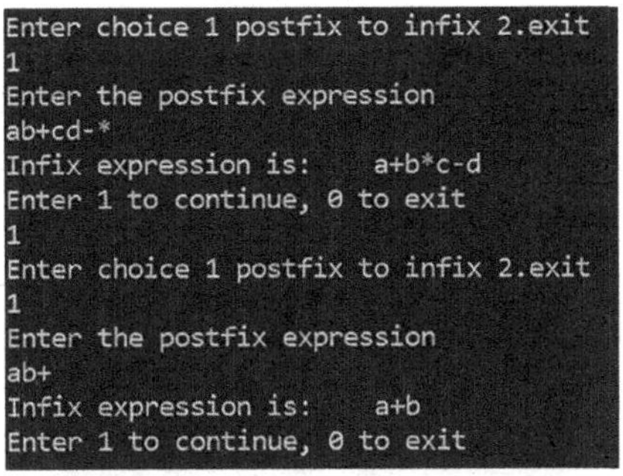

```
Enter choice 1 postfix to infix 2.exit
1
Enter the postfix expression
ab+cd-*
Infix expression is:    a+b*c-d
Enter 1 to continue, 0 to exit
1
Enter choice 1 postfix to infix 2.exit
1
Enter the postfix expression
ab+
Infix expression is:    a+b
Enter 1 to continue, 0 to exit
```

2.4.2 Implementation of the Tower of Hanoi Puzzle Using Recursion

2.4.2.1 Recursive Algorithm of the Tower of Hanoi Puzzle

Step 1: Start

Step 2: Move n-1 disks from source rod (from_rod) to auxiliary rod (aux_rod) using destination rod (to_rod)

Step 3: Move n^{th} disk from source rod to destination rod

Step 4: Move n-1 disks from auxiliary rod to destination rod using source rod

Step 5: Stop

In the above algorithm, from_rod refers to the source rod, to_rod refers to the destination rod and aux_rod refers to the auxiliary rod.

2.4.2.2 Program of Tower of Hanoi Puzzle Using Recursion

```
#include <stdio.h>
void towerOfHanoi(int n, char from_rod, char to_rod, char
aux_rod)
{
      if (n == 1)
      {
                printf("\n Move disk 1 from rod %c to rod %c",
                from_rod, to_rod);
                return;
      }
      towerOfHanoi(n-1, from_rod, aux_rod, to_rod);
      printf("\n Move disk %d from rod %c to rod %c", n, from_
      rod, to_rod);
      towerOfHanoi(n-1, aux_rod, to_rod, from_rod);
}

int main()
{
      int n = 3;  // Number of disks
      towerOfHanoi(n, 'A', 'C', 'B');  // A, B and C are names
      of rods
      return 0;
}
```

Output:

```
Move disk 1 from rod A to rod C
Move disk 2 from rod A to rod B
Move disk 1 from rod C to rod B
Move disk 3 from rod A to rod C
Move disk 1 from rod B to rod A
Move disk 2 from rod B to rod C
Move disk 1 from rod A to rod C
```

Note that if the discs are three, then total moves are ($2^3 - 1$), i.e., 7. That is, if there are n discs, total ($2^n - 1$) moves are required to solve this puzzle.

2.5 Queue

Queue is **non-primitive linear data structure.** Queue is a data structure that allows us to use two ends of it named as **front and rear**. In the queue data structure, from the front, one can remove elements, and from the rear, one can insert elements. Queue is also called as first in first out (FIFO). A queue is a useful data structure for programming. It is similar to the ticket queue outside a cinema, where the first person in the queue is the first person who gets the ticket.

Types of Queues

There are four types of queues, as follows:

1. **Linear queue:** In a linear queue, the insertion takes place in the rear and the withdrawal takes place at the front. It is strictly in accordance with the FIFO rule.
2. **Circular queue:** In a circular queue, the final item points to the first item with a circular link. The main advantage of a circular queue compared to a linear queue is better memory usage. If the last position is full and the first position is empty, it is possible to insert an item into the first position. You cannot do this in a linear queue.
3. **Priority queue:** In the priority queue, items are ordered by key value so that item with the minimum value of the key is in front, and item with the highest value of the key is at the rear or vice versa. A priority queue is a particular type of queue in which each item is associated with a priority and is served based on its priority. When items with the same priority take place, they are executed in their order in the queue. The insertion of elements into the priority queue occurs according to the arrival of the values and the removal of the elements occurs according to their priority.
4. **Double-ended queue:** Double-ended queue is also referred to as dequeue. In a double-ended queue, insertion and deletion operations are carried out from both the front and rear ends.

2.5.1 Linear Queue

2.5.1.1 Representation and Operations on Linear Queue Using Arrays

All elements of the linear queue are stored sequentially. We can simply represent linear queue byww using linear arrays. There are two types of indices: front index and rear index. Front and rear index point to the position from where insertions and deletions are performed in a linear queue. Initially, the value of the front and rear index in a linear queue is –1, which represents an empty queue. Array representation of a linear queue containing six elements along with the respective values of front and rear indices is shown in Figure 2.7.

Various operations on the linear queue are as follows:

1. Linear queue overflow operation.
2. Linear queue underflow operation.
3. Inserting the element into the linear queue is also called an enqueue operation.
4. Removing an element from the linear queue is also called a dequeue operation.
5. Display the elements of the linear queue.

1. **Linear queue overflow operation:**

   Queue overflow results from trying to add an element into a full queue.
2. **Linear queue underflow operation:**

   Queue underflow occurs when you try to delete an item in an empty queue.
3. **Inserting the element into the linear queue (Figure 2.8):**

   Before performing the insertion operation, you must verify if the queue is full or not. If the rear pointer exceeds the maximum length of the queue, the queue overflow occurs. If the queue does not overflow, then insert the element from the rear end.

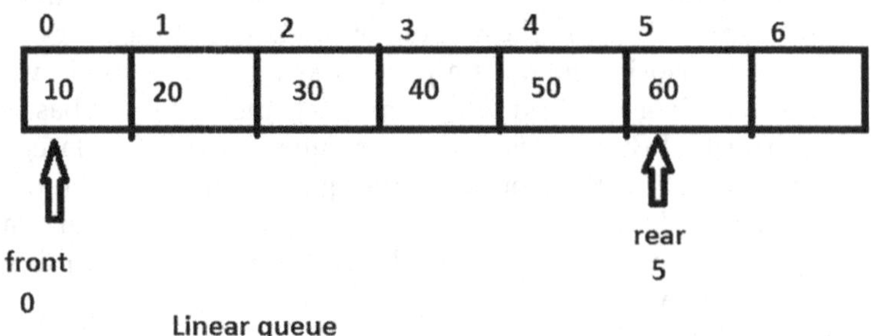

FIGURE 2.7
Linear queue representation using an array.

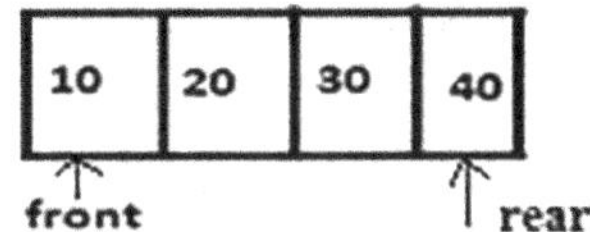

FIGURE 2.8
Insertion of element into the linear queue.

4. **Removing an element from the linear queue:**
   Removing any item from the queue only starts from the front. Before performing any delete operation, one must check whether the queue is empty or not. If there is an empty queue, deletion is not possible (Figure 2.9).
   While removing the element from the linear queue which is implemented using arrays, front index is incremented that is an element is removed from the queue logically, but the physical element is present at the same index location in linear queue. In the above figure, while deleting the element from linear queue, the only front index gets incremented from index 0 to 1, but physical element 10 remains at the same index location and linear queue is being treated from front to rear only, that is from index 1 to rear index 3.
5. **Display the elements of the linear queue:**
   If the linear queue is not empty, then elements are displayed from the front end to the rear end while performing display operation on it.

2.5.1.2 Implementation of Linear Queue Using Arrays

```
#include<stdio.h>
#include<stdlib.h>
#define max 5

// Linear queue implemented using array
struct queue
{
 int que[max];
 int front,rear;
}q;
```

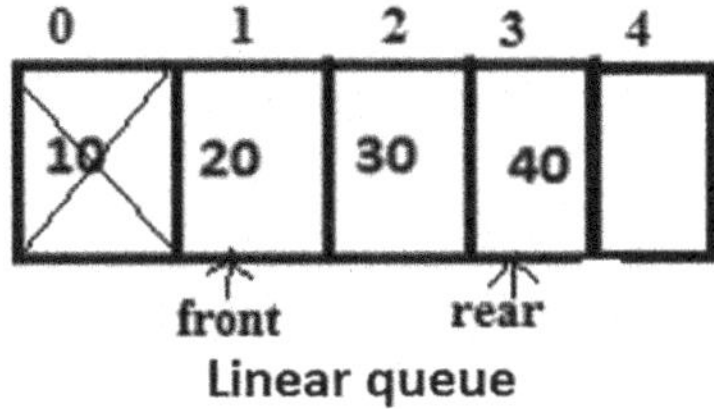

FIGURE 2.9
Removing an element from the linear queue.

```
// Verify that the linear queue is empty or not
int qempty()
{
if((q.front==-1)||(q.front>q.rear))
{
return 1;
}
else
return 0;
}

// Verify that the linear queue is full or not
int qfull()
{
if(q.rear>=max-1)
return 1;
else
return 0;
}

// Inserting the element in the linear queue
void qinsert(int data)
{
if(q.front==-1)
{
q.front++;
}
q.que[++q.rear]=data;
}

// Remove an item from the linear queue
void qdelete()
{
int data;
data=q.que[q.front];
q.front++;
printf("Deleted element=%d\n", data);
}

// Display the elements from the linear queue
void display()
{
int i;
i=q.front;
printf("Queue elements are : ");
for(; i<=q.rear; i++)
{
printf("\t%d",q.que[i]);
}
printf("\n");
}
```

```
// main function definition
int main()
{
 int choice, data;
 char ch;
 q.front=-1;
 q.rear=-1;
 do
 {
  printf(" Linear queue implementation using array: ");
  printf("\n 1.insert \t 2.delete \t 3.display \t 4.exit");
  printf("\n Enter choice: ");
  scanf("%d", &choice);
  switch(choice)
  {
case 1:
if(qfull())
{
printf("\n Queue is full");
}
else
{
printf(" Enter data: ");
scanf("%d",&data);
qinsert(data);
}
break;
case 2:
if(qempty())
printf("\n Queue is empty");
else
qdelete();
break;
case 3:
if(qempty())
printf("\n Queue is empty");
else
display();
break;
case 4:
exit(0);
default:
printf("\n wrong choice");
}
printf("Do you want to continue...[y/n]");
// fflush function clears the standard input device buffer
fflush(stdin);
scanf("%c",&ch);
}while(ch=='y');
return 0;
}
```

Output:

```
 Linear queue implementation using array:
 1.insert        2.delete        3.display       4.exit
 Enter choice: 1
 Enter data: 21
Do you want to continue...[y/n]y
 Linear queue implementation using array:
 1.insert        2.delete        3.display       4.exit
 Enter choice: 1
 Enter data: 51
Do you want to continue...[y/n]y
 Linear queue implementation using array:
 1.insert        2.delete        3.display       4.exit
 Enter choice: 1
 Enter data: 91
Do you want to continue...[y/n]y
 Linear queue implementation using array:
 1.insert        2.delete        3.display       4.exit
 Enter choice: 1
 Enter data: 47
Do you want to continue...[y/n]y
 Linear queue implementation using array:
 1.insert        2.delete        3.display       4.exit
 Enter choice: 3
Queue elements are :    21      51      91      47
Do you want to continue...[y/n]y
 Linear queue implementation using array:
 1.insert        2.delete        3.display       4.exit
 Enter choice: 2
Deleted element=21
Do you want to continue...[y/n]y
 Linear queue implementation using array:
 1.insert        2.delete        3.display       4.exit
 Enter choice: 3
Queue elements are :    51      91      47
Do you want to continue...[y/n]n
```

2.5.2 Circular Queue

In the case of linear queue using array, the elements are deleted logically, and this can be shown as follows (Figure 2.10):

We have deleted elements 10, 20 and 30 means simply front index of a linear queue is shifted ahead. We will consider a linear queue from front to rear

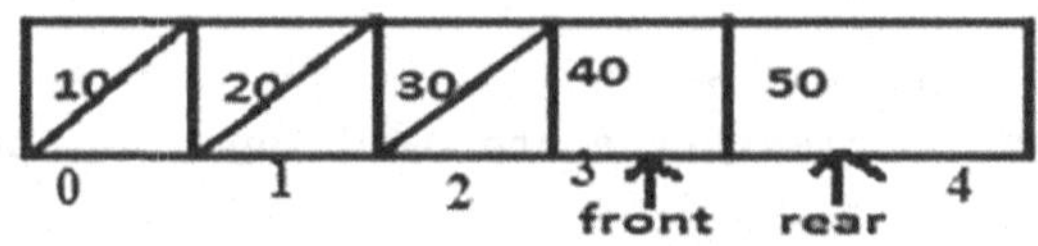

FIGURE 2.10
Deletion operation in a linear queue.

only. Here in the above linear queue, if we try to insert any more elements, then it won't be possible as it will give "queue full" message. Although there is space for elements 10, 20, 30, which is already deleted, we cannot utilize this space because the type of queue is a linear queue.

Hence, to overcome the drawback of the linear queue, we will introduce another type of queue called as a circular queue. The main advantage of the circular queue is that we can utilize the space of a circular queue fully.

2.5.2.1 Representation and Operations on Circular Queue Using Arrays

A circular queue is represented similarly to a linear queue except that the last position is linked to the first position in a circular queue that forms a circle, which is shown in Figure 2.11:

Consider the deleted items such as 10, 50 and 40 in the above circular queue. Then the forehead is incremented and pointed toward index 3. The places in index 0, 1 and 2 are now available to insert a new item, and this space is then reused in the circular queue, which is the linear queue limitation. There is a formula that has to be applied to set the front and rear indexes in a circular queue, which is implemented using arrays as shown below:

In circular queue implementation using array, rear and front index is calculated as

```
rear = (rear+1)%size
front = (front+1)%size
```

For example, in Figure 2.11, the size of the circular queue is 5, if rear points to 4 index, and front points, to the index 3, then index 0, 1 and 2 places are empty in the circular queue, and if we insert a new element in the circular queue, then from rear end, we will insert it. In this situation, the rear is calculated as

```
rear = (rear + 1) % size
       = (4 + 1) % 5
rear = 0
```

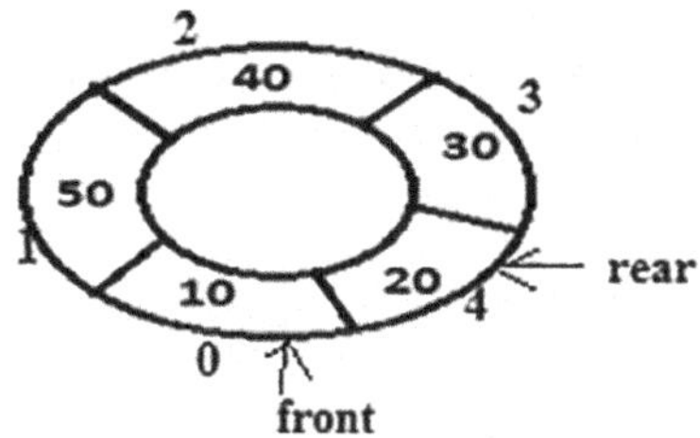

FIGURE 2.11
Circular queue.

Then, a new element is inserted into the circular queue at the rear index 0.

Let us consider that front is pointing to index 3. Similarly, while deleting the element from the circular queue, front is calculated as,

```
front  = (front + 1) % size
= (3 + 1) % 5
front = 4
```

Therefore, the element at the third position is deleted and the front gets incremented to the fourth index and the front points to the fourth position element in the circular queue.

Various operations on the circular queue are as follows:

1. Overflow operation
2. Underflow operation
3. Inserting the element into the circular queue is also called as enqueue operation.
4. Removing an element from the circular queue is also called as dequeue operation.
5. Display the elements of the circular queue.

Some of the important conditions we must know about the circular queue are as follows:

1. If the rear and front points to the same index, then circular queue consists of only one element.
2. If the front index is greater than one than the rear index, which means if the front index is 4 and the rear index is 3, the circular queue is full.

   If (front == (rear+1) % max size of the circular queue), if this condition is true, then and then only circular queue is full.
3. If front and rear index point to the –1, then the circular queue is said to be empty (Figure 2.12).

In the above circular queue, if we want to insert a new item into the circular queue, then increment the rear index as follows.

The current value of the rear index is 4 and the size of the circular queue is 5, and then the rear index of the newly inserted element is computed as,

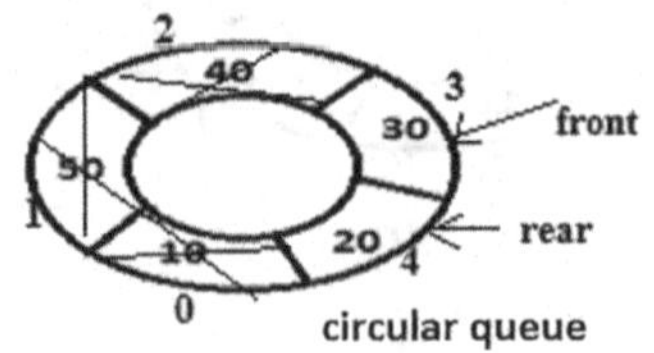

FIGURE 2.12
Enqueue and Dequeue operation on circular queue.

```
rear = (rear + 1) % SIZE     ( SIZE = 5, which corresponds to
the size of the circular queue)
rear = (4 + 1) % 5
rear = 0
```

Therefore, insert the element at the rear index 0 in our circular queue. The circular queue is not full because front is pointing to index 3. In this way, we can utilize the deleted space in the circular queue.

2.5.2.2 Implementation of Circular Queue Using Arrays

```
#include<stdio.h>
#include<stdlib.h>
#define max 5

// Circular queue using array
int que[max];
int front,rear;

// Check whether circular queue is full or not
int qfull()
{
if(front==(rear+1)%max)
return 1;
else
return 0;
}

// Check whether circular queue is empty or not
int qempty()
{
if(front==-1)
return 1;
else
return 0;
}

// Insertion of element into circular queue
void insert(int data)
{
if(front==-1)
{
front=rear=0;
que[rear]=data;
}
else
{
rear=(rear+1)%max;
que[rear]=data;
}
}
```

```
// Delete an element from the circular queue
void delete()
{
int data;
if(front==rear)
{
data=que[front];
front=rear=-1;
}
else
{
data=que[front];
front=(front+1)%max;
}
printf("Deleted element=%d\n", data);
}

// Display the elements from the circular queue
void display()
{
int i=front;
printf("Elements in circular queue are: ");
do
{
printf("\t%d",que[i]);
i=(i+1)%max;
}while(i != ((rear + 1)%max));
printf("\n");
}

// main function
int main()
{
 int choice, data;
 char ch;
 front=-1;
 rear=-1;
 do
 {
printf(" Circular Queue:");
printf(" 1:insert\t 2:delete\t 3:display\t 4.exit\n");
printf(" Enter your choice:");
scanf("%d", &choice);
switch(choice)
{
case 1:
printf("Enter data: ");
scanf("%d",&data);
if(qfull())
printf("\nqueue is full");
else
insert(data);
```

```
break;
case 2:
if(qempty())
printf("\n Circular queue is empty");
else
delete();
break;
case 3:
if(qempty())
printf("\n Circular queue is empty");
else
display();
break;
case 4:
exit(0);
default:
printf("\n Wrong choice");
}
printf(" Do you want to continue...[y/n]: ");
fflush(stdin);
scanf("%c",&ch);
}while(ch=='y');
return 0;
}
```

Output:

```
 Circular Queue: 1:insert        2:delete        3:display       4.exit
 Enter your choice:1
Enter data: 44
 Do you want to continue...[y/n]: y
 Circular Queue: 1:insert        2:delete        3:display       4.exit
 Enter your choice:22

 Wrong choice Do you want to continue...[y/n]: y
 Circular Queue: 1:insert        2:delete        3:display       4.exit
 Enter your choice:1
Enter data: 88
 Do you want to continue...[y/n]: y
 Circular Queue: 1:insert        2:delete        3:display       4.exit
 Enter your choice:1
Enter data: 99
 Do you want to continue...[y/n]: y
 Circular Queue: 1:insert        2:delete        3:display       4.exit
 Enter your choice:1
Enter data: 11
 Do you want to continue...[y/n]: y
 Circular Queue: 1:insert        2:delete        3:display       4.exit
 Enter your choice:3
Elements in circular queue are:         44      88      99      11
 Do you want to continue...[y/n]: y
 Circular Queue: 1:insert        2:delete        3:display       4.exit
 Enter your choice:2
Deleted element=44
 Do you want to continue...[y/n]: y
 Circular Queue: 1:insert        2:delete        3:display       4.exit
 Enter your choice:3
Elements in circular queue are:         88      99      11
 Do you want to continue...[y/n]: y
 Circular Queue: 1:insert        2:delete        3:display       4.exit
 Enter your choice:4
```

2.5.3 Priority Queue

Restriction on stack and queue is that elements should be inserted or deleted from a specific end, the sequence of elements is ignored in the stack or queue operation. Elements in the priority queue should have some specific ordering. In a priority queue, high-priority items are served before low-priority items. A typical example of this priority queue is scheduling the jobs in the operating system.

2.5.3.1 Types of Priority Queues

1. **Ascending order priority queue:** It is a collection of items in which items can be inserted randomly, but the only smallest element can be removed first from the ascending priority queue.
2. **Descending order priority queue:** It is a collection of items in which items can be inserted randomly, but the only largest element can be removed first from the descending priority queue.

   When we try to implement a priority queue using arrays, then insertion becomes very simple, but for the deletion of any element from the priority queue raises two issues:

   1. Either examine all elements of the priority queue or find out the largest or smallest element for deletion.
   2. Deletion of the item in the middle of an array is a difficult task that requires shifting of the elements in an array.

   These issues are addressed in the following ways:

   1. In array, if any arbitrary element is deleted, then this place becomes empty. Therefore, a special indicator such as –1 has to be placed in that position which indicates the element is deleted from that position.
   2. There should be shifting of elements after each deletion to manage contiguous storage in an array.
   3. For each deletion of an element from the priority queue scanning of the entire priority queue is required. Here the solution is that you insert numbers arbitrarily, and after each insertion, the ordering will be maintained in the priority queue. The elements can be assumed either in ascending or in descending order. Hence, there will not be a need to scan the entire list for deletion of the largest or smallest element. Insertion will be from the rear end, and deletion will be from the front end only in a given priority queue.

2.5.3.2 Implementing Ascending Order Priority Queue Using Arrays

```
#include<stdio.h>
#include<stdlib.h>
#define max 100
int que[max];
int front, rear;
```

```
// Check the priority queue is empty or not
int qempty()
{
if((front==-1)||(front>rear))
return 1;
else
return 0;
}

// Check the priority queue is full or not
int qfull()
{
if(rear>=max-1)
return 1;
else
return 0;
}
void priority();

// Insert an element in priority queue
void insert(int data)
{
if(front==-1)
{
front++;
}
que[++rear]=data;
priority();
}

// Check the priority of an element and arrange the data in
ascending order
void priority()
{
int i,j,temp;
for(i=front;i<rear;i++)
{
for(j=front;j<rear;j++)
{
if(que[j]>que[j+1])
{
temp=que[j];
que[j]=que[j+1];
que[j+1]=temp;
}
}
}
}

// Delete the element from the priority queue
void delete()
```

```
{
int data;
data=que[front];
front++;
printf(" Deleted element=%d \n", data);
}
//display element from priority queue
void display()
{
int i;
i=front;
printf("Elements in priority queue: ");
while(i<=rear)
{
printf("\t%d",que[i]);
i++;
}
printf("\n");
}
int main()
{
 int choice, data;
 char ch;
 front=-1;
 rear=-1;
 do
 {
  printf(" Priority Queue");
  printf(" 1:insert\t2:delete\t3:display\t4:exit");
  printf("\n Enter your choice");
  scanf("%d", &choice);
  switch(choice)
  {
   case 1:
printf("Enter data: ");
      scanf("%d",&data);
      if(qfull())
      printf("\n Queue is full");
      else
      insert(data);
   break;
   case 2:
if(qempty())
      printf("\n Queue is empty");
      else
      delete();
   break;
   case 3:
if(qempty())
      printf("\n Queue is empty");
      else
```

```
        display();
    break;
    case 4:
exit(0);
    default:
printf("\n Wrong choice");
   }
printf(" Do you want to continue[y/n]");
fflush(stdin);
scanf("%c",&ch);
}while(ch=='y');
return 0;
}
```

Output:

```
 Priority Queue 1:insert       2:delete        3:display       4:exit
 Enter your choice1
Enter data: 21
 Do you want to continue[y/n]y
 Priority Queue 1:insert       2:delete        3:display       4:exit
 Enter your choice1
Enter data: 31
 Do you want to continue[y/n]y
 Priority Queue 1:insert       2:delete        3:display       4:exit
 Enter your choice1
Enter data: 41
 Do you want to continue[y/n]y
 Priority Queue 1:insert       2:delete        3:display       4:exit
 Enter your choice1
Enter data: 51
 Do you want to continue[y/n]y
 Priority Queue 1:insert       2:delete        3:display       4:exit
 Enter your choice3
Elements in priority queue:    21      31      41      51
 Do you want to continue[y/n]y
 Priority Queue 1:insert       2:delete        3:display       4:exit
 Enter your choice2
 Deleted element=21
 Do you want to continue[y/n]y
 Priority Queue 1:insert       2:delete        3:display       4:exit
 Enter your choice3
Elements in priority queue:    31      41      51
 Do you want to continue[y/n]y
 Priority Queue 1:insert       2:delete        3:display       4:exit
 Enter your choice4
```

2.5.4 Double-ended Queue

Double-ended Queue is a special type of queue data structure in which insert and delete operations are performed at both front and rear ends. This means that we can enter the element from both the front and rear positions and can remove the element from both front and rear positions.

2.5.4.1 Representation of Double End Queue

Double-ended queues can be represented in two different ways, and these are as follows:

1. **Input-restricted double-ended queue:**
   The input-restricted double-ended queue means that some constraints apply to the insertion of the elements. In an input-restricted double-ended queue, the insertion operation is carried out only from the rear end, and deletion operation is carried out at both the rear and front ends (Figure 2.13).

2. **Output-restricted double-ended queue:**
   The output-restricted double-ended queue means that some constraints apply to the deletion of the elements. In an output-restricted double-ended queue, the deletion activity is carried out only from the front end, and insertion operation is carried out at both the rear and front ends (Figure 2.14).

2.5.4.2 Operations Performed on the Double-Ended Queue

Various operations performed on the double-ended queue are as follows:

1. Insert at the front end
2. Delete from front end
3. Insert at the rear end
4. Delete from the rear end

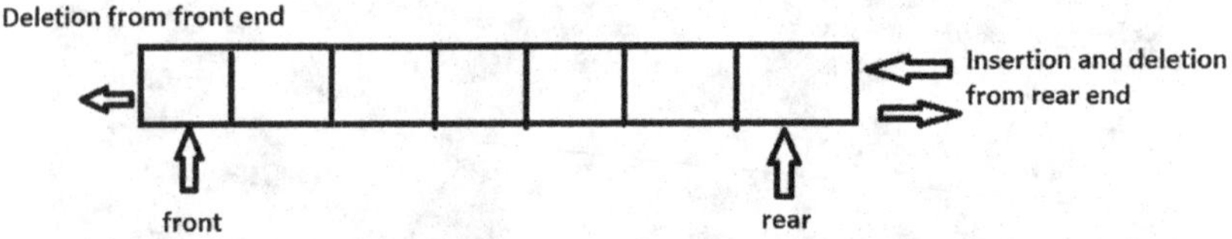

FIGURE 2.13
Input-restricted double-ended queue representation.

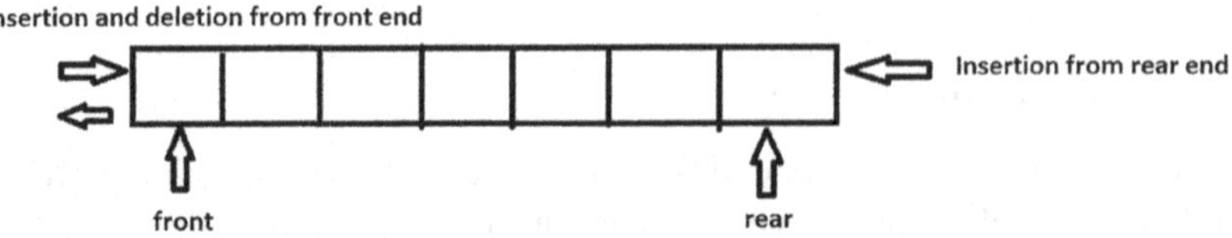

FIGURE 2.14
Output-restricted double-ended queue representation.

5. Double-ended queue full operation
6. Double-ended queue empty operation
7. Display the elements of the double-ended queue

2.5.4.3 Implementing Double Ended Queue Using Arrays

```
#include<stdio.h>
#define SIZE 21

//The various operations performed on the Double Ended queue
int deQueueFront(); // Delete from front end
int deQueueRear();  // Delete from rear end
void enQueueRear(int); // Insert at rear end
void enQueueFront(int); // Insert at front end
void display();  // Display the elements
int queue[SIZE];
int rear = 0, front = 0;

// main function definition
int main()
{
    char ch;
    int choice1, choice2, value;
    printf("Double Ended Queue: \n");
     do
     {
          printf("\n1.Input Restricted Double Ended Queue:
\n");
          printf("2.Input Restricted Double Ended Queue: \n");
          printf("\nEnter your choice: ");
          scanf("%d",&choice1);
          switch(choice1)
          {
               case 1:
                    printf("\n Select the Operation: \n");
                    printf("1.Insert from Front\n2.Delete from
Rear\n3.Delete from Front\n4. Display");
                    do
                    {
                       printf("\nEnter your choice for the
operation: ");
                       scanf("%d",&choice2);
                       switch(choice2)
                       {
                          case 1: enQueueRear(value);
                                  display();
                                 break;
                            case 2: value = deQueueRear();
                             printf("\nThe value deleted from
rear end of Dequeue is %d",value);
```

```
                              display();
                            break;
                      case 3: value=deQueueFront();
                              printf("\nThe value deleted
from front end of Dequeue is %d",value);
                              display();
                               break;
                      case 4: display();
                              break;
                      default:printf("Wrong choice");
                   }
                   printf("\nDo you want to continue
another operation (Y/N): ");
                   ch=getche();
                }while(ch=='y'||ch=='Y');
                getch();
                break;

          case 2 :
                printf("\nSelect the Operation:\n");
                printf("1. Insert at Rear\n2. Insert at
Front\n3. Delete from Front\n4. Display");
                do
                {
                   printf("\nEnter your choice for the
operation: ");
                   scanf("%d",&choice2);
                   switch(choice2)
                   {
                      case 1: enQueueRear(value);
                              display();
                              break;
                      case 2: enQueueFront(value);
                              display();
                              break;
                      case 3: value = deQueueFront();
                              printf("\nThe value deleted
from front end of Dequeue is %d",value);
                              display();
                              break;
                      case 4: display();
                              break;
                      default:printf("Wrong choice");
                   }
                   printf("\nDo you want to continue
another operation (Y/N): ");
                   ch=getche();
                } while(ch=='y'||ch=='Y');
                getch();
                break ;
```

```
            }
            printf("\nDo you want to continue (Y/N):");
            ch=getch();
      }while(ch=='y'||ch=='Y');
}

 // Insert at rear end
void enQueueRear(int data)
{
     char ch;
     if(front == SIZE/2)
      {
            printf("\n Queue is full. ");
            return;
      }
      do
      {
            printf("\nEnter the value to be inserted:");
            scanf("%d",&data);
            queue[front] = data;
            front++;
            printf("Do you want to continue insertion (Y/N)
\n");
            ch=getche();
      }while(ch=='y');
}

// Insert at front end
void enQueueFront(int data)
{
     char ch;
     if(front==SIZE/2)
      {
            printf("\nQueue is full.");
            return;
      }
      do
      {
            printf("\nEnter the value to be inserted:");
            scanf("%d",&data);
            rear--;
            queue[rear] = data;
            printf("Do you want to continue insertion (Y/N");
            ch = getche();
      }
      while(ch == 'y');
}
```

```
// Delete from rear end
int deQueueRear()
{
     int deleted;
     if(front == rear)
     {
           printf("\nQueue is Empty.");
           return 0;
     }
     front--;
     deleted = queue[front+1];
     return deleted;
}

// Delete from front end
int deQueueFront()
{
     int deleted;
     if(front == rear)
     {
           printf("\nQueue is Empty. ");
           return 0;
     }
     rear++;
     deleted = queue[rear-1];
     return deleted;
}

// Display the elements
void display()
{
     int i;
     if(front == rear)
        printf("\nQueue is Empty. ");
     else{
        printf("\nThe Queue elements are:");
        for(i=rear; i < front; i++)
        {
           printf("%d\t ",queue[i]);
        }
     }
}
```

Output:

```
Double Ended Queue:

1.Input Restricted Double Ended Queue:
2.Input Restricted Double Ended Queue:

Enter your choice: 1

Select the Operation:
1.Insert from Front
2.Delete from Rear
3.Delete from Front
4. Display
Enter your choice for the operation: 1

Enter the value to be inserted:21
Do you want to continue insertion (Y/N)
y
Enter the value to be inserted:31
Do you want to continue insertion (Y/N)
y
Enter the value to be inserted:41
Do you want to continue insertion (Y/N)
n
The Queue elements are:21       31      41
Do you want to continue another operation (Y/N): y
Enter your choice for the operation: 3

The value deleted from front end of Dequeue is 21
The Queue elements are:31       41
Do you want to continue another operation (Y/N): n
Do you want to continue (Y/N):
```

2.6 Applications of the Queue Data Structure

1. The queue is useful in CPU scheduling and disk scheduling. When multiple processes require CPU at once, various CPU scheduling algorithms are used that are implemented using the queue data structure.
2. When data is transferred asynchronously from one process to the next, the queue is used for synchronization, for example, IO Buffers, pipes, file IO, etc.
3. In the print spooling, documents are loaded into a buffer, and then the printer pulls them off the buffer at its own rate. Spooling also lets you place a number of print jobs on a queue, on the other hand, of waiting for each one to complete before specifying the next one.

4. While implementing breadth first search (BFS) algorithm on a nonlinear graph data structure requires linear queue data structure. It is an algorithm for traversing or searching in nonlinear graph data structures. It initiates at some random node of a graph and explores the neighbor nodes first, before moving to the next level neighbors.
5. Handling of interrupts in real-time systems. The interrupts are tackled in the same order as they appear, first come, first served using the queue.
6. In real life, call center telephone systems will use queues, to hold people calling them in an order, until a service representative is free.
7. The implementation of the prim algorithm can be achieved through priority queues.
8. The implementation of the shortest route algorithm of Dijkstra can be done using priority queues.
9. Prioritized queues are used in the operating system for load balancing and interrupt handling.
10. Priority queues are used in Huffman codes for data compression.
11. Queues are used in traffic systems as well.
12. The queue data structure is used in the history of web browsers. More recently, visited URLs are added to the front of the deque, and the URL on the back of the deque is removed after some stipulated number of insertions at the front.
13. Some other ordinary application of the deque is storing a software application's list of undo operations.
14. In the network system, queues are used in routers and switches.
15. In e-mails as well, the queue data structure is used.

2.7 Differences between Stack and Queue Data Structure

Stack	Queue
The stack is a non-primitive linear data structure.	The queue is also a non-primitive linear data structure.
The stack is implemented using basic data structures such as an array or linked list.	The queue is also implemented using basic data structures such as an array or linked list.
The stack is a linear data structure in which elements are inserted and deleted from only one end called top of the stack and is based on the LIFO principle.	The queue is a linear data structure in which elements are inserted from one end called a rear end and removed from the other end called the front end and is based on the FIFO principle.

(*Continued*)

Stack	Queue
Adding and removing of elements are carried out from only one end in the stack.	Adding and removing of elements are carried out from different ends in the queue.
Inserting an element on the stack is called push operation.	Insert operation in the queue is called enqueue.
Deleting an element from the stack is called pop operation.	Remove element operation in the queue is called as dequeue operation.
In the Stack data structure, only one pointer is maintained, which is pointing to the top element of the stack.	The queue data structure needs two pointers in their implementation, pointing to the rear end and pointing to the front end.
The stack data structure is an iterative data structure and is used to resolve problems based on recursion.	The queue is a sequential data structure and is used for troubleshooting issues involving sequential processing.
The stack can be used to solve issues like pre-order, post-order and in-order traversal of a binary tree.	Producer consumer problems are addressed using a queue data structure as the underlying data structure.
Stack data structure can be considered as a vertical linear collection.	The queue data structure can be considered as a horizontal linear collection.
The collection of plates placed above each other is a specific real-world example of the stack.	People standing in a queue to pay the electricity bill or take a ticket from the movie counter are the specific real-world example of a queue.

2.8 Interview Questions

1. What is a queue?
2. What is a priority queue?
3. What are the disadvantages of sequential storage?
4. What are the disadvantages of representing a stack or queue by a linked list?
5. List out the areas in which data structures are applied extensively.
6. What is LIFO?
7. Which data structures are applied when dealing with a recursive function?
8. What is a stack?
9. What is the difference between a PUSH and a POP operation?
10. List different applications of queue data structure.
11. Explain what are Infix, Prefix and Postfix Expressions?
12. Why and when should we use stack and queue data structures instead of arrays or Lists?
13. What are different operations we can perform on queues?

14. What is the advantage of the heap over a stack?
15. What is a dequeue?
16. What is the state of the stack after the following sequence of pushes and pops?

```
 Stack<int> s;
s.push( 3 );
s.push( 5 );
s.push( 2 );
s.push( 15 );
s.push( 42 );
s.pop();
s.pop();
s.push( 14 );
s.push( 7 );
s.pop();
s.push( 9 );
s.pop();
s.pop();
s.push( 51 );
s.pop();
s.pop();
```

17. Suppose the numbers 0, 1, 2,..., 9 were pushed onto a stack in that order but that pops to occur at random points between the various pushes. The following is a valid sequence in which the values in the stack could have been popped:

    3, 2, 6, 5, 7, 4, 1, 0, 9, 8

    Explain why it is not possible that

    3, 2, 6, 4, 7, 5, 1, 0, 9, 8 is a valid sequence in which the values could have been popped off the stack.
18. Consider an empty stack of integers. Let the numbers 1, 2, 3, 4, 5 and 6 be pushed onto this stack only in the order they appeared from left to right. Let S indicate a push and X indicate a pop operation. Can they be permuted into order 325641 (output) and order 154623? (if a permutation is possible, give the order string of operations).

    (Hint: SSSSSSXXXXXX outputs 654321)
19. How to implement two stacks using only one array. Your stack practices must not indicate an overflow unless every slot in the array is used?
20. How to implement stack using two queues?

2.9 Multiple Choice Questions

1. Stack data structure is based on which of the following principle
   A. First in first out
   B. First in last out
   C. Last in last out
   D. Last in first out

   Answer: (D)

2. Which of the following is not the type of queue data structure?
   A. Priority queue
   B. Circular queue
   C. Single-ended queue
   D. Ordinary queue

   Answer: (C)

3. Which of the following data structure is non-primitive linear data structure?
   A. Stack
   B. Graph
   C. Trees
   D. Binary tree

   Answer: (A)

4. Which of the below is true about stack implementation using a linked list?
   A. In a push operation, if new nodes are inserted at the beginning of the linked list, then in pop operation, nodes must be deleted from the end.
   B. In a push operation, if new nodes are inserted at the end of the linked list, then in pop operation, nodes must be deleted from the beginning.
   C. Both of the above are true.
   D. None of the above is true.

   Answer: (D)

5. Assume that a stack is to be implemented with a linked list rather than an array. What is the time complexity of the push and pop operations of the stack implemented using a linked list hoping that stack is implemented effectively?
   A. O(n log k)
   B. O(nk)
   C. $O(n^2)$
   D. $O(k^2)$

   Answer: (A)

6. Which of the following statements is false or true?
   A. The stack is a non-primitive linear data structure.
   B. The queue is a primitive nonlinear data structure.
      A. Statement 1 is false
      B. **Statement 2 is false**
      C. Statements 1 and 2 are false
      D. Statements 1 and 2 are true

   Answer: (B)
   Explanation:
   The queue is a non-primitive linear data structure.

7. The seven elements A, B, C, D, E, F and G are pushed onto a stack in reverse order, that is, starting from G. The stack is popped five times, and each element is inserted into a queue. Two elements are deleted from the queue and pushed back onto the stack. Now, one element is popped from the stack. The popped item is ________.
   A. A
   B. B
   C. F
   D. D. G

   Answer: (B)

8. Which of the following is an application of non-primitive linear queue data structure?
   A. When a resource is shared among multiple consumers?
   B. When data is transferred asynchronously between two processes?
   C. Load balancing
   D. All of the above

   Answer: (D)

9. Which of the following statements is false or true?
   A. The queue is useful in CPU scheduling, disk scheduling.
   B. Priority queues are used in Huffman codes for data compression.
      A. Statement 1 is false
      B. Statement 2 is false
      C. Statements 1 and 2 are false
      D. Statements 1 and 2 are true

   Answer: (D)

10. Which of the following statements is false or true?
    A. In the DFS graph, the traversing technique queue data structure is used.
    B. In expression evaluation, the stack data structure is used.
       A. Statement 1 is false
       B. Statement 2 is false
       C. Statements 1 and 2 are false
       D. Statements 1 and 2 are true

    Answer: (A)
    Explanation:
    In the DFS graph, the traversing technique stack data structure is used.

11. Stack cannot be used to
    A. Implement recursion
    B. Arithmetic expression evaluation
    C. Parsing or syntax analysis
    D. In CPU scheduling and disk scheduling.

    Answer: (D)
    Explanation: Stack can be used to solve arithmetic expression, implementing recursion and parsing. While queue is useful in CPU scheduling and disk scheduling.

12. A single array A[1..MAXSIZE] is used to implement two stacks. The two stacks grow from opposite ends of the array. Variables top1 and top2 (topl< top 2) point to the location of the topmost element in each of the stacks. If the space is to be used efficiently, the condition for "stack full" is (GATE CS 2004)
    A. (top1 = MAXSIZE/2) and (top2 = MAXSIZE/2+1)
    B. top1 + top2 = MAXSIZE

C. (top1= MAXSIZE/2) or (top2 = MAXSIZE)

D. top1= top2 -1

Answer: (D)

Explanation

If we want to use space efficiently, then the size of any stack can be more than MAXSIZE/2. Both stacks will grow from both ends, and if any of the stack top reaches near the other top, then stacks are full. Therefore, the condition will be top1 = top2 -1 (given that top1 < top2)

13. Which one of the following is an application of stack data structure?

A. Implement recursion

B. Towers of Hanoi problems

C. Arithmetic expression evaluation

D. All of the above

Answer: (D)

14. The postfix form of the expression (A+ B)*(C*D- E)*F / G is?

A. AB+ CD*E - FG /**

B. AB + CD* E - F **G /

C. AB + CD* E - *F *G /

D. AB + CDE * - * F *G /

Answer: (A)

15. The non-primitive linear data structure required to check whether an expression contains balanced parenthesis is

A. Stack

B. Queue

C. Array

D. Tree

Answer: (A)

16. If the elements "A", "B", "C" and "D" are placed in a stack and are removed one at a time, in what order will they be deleted?

A. ABCD

B. DCBA

C. DCAB

D. ABDC

Answer: (B)

17. What is the minimum number of stacks of size n required to implement a queue data structure of size n?

    A. One

    B. Two

    C. Three

    D. Four

    Answer: (B)

18. Consider the following array implementation of stack:

```
#define MAX 10
Struct STACK
{
Int arr [MAX];
Int top = -1;
}
```

If the array index starts with 0, the maximum value of top that does not cause stack overflow is?

    A. 8

    B. 9

    C. 10

    D. 11

    Answer: (A)

19. Which of the following applications utilizes a stack data structure?

    A. Parenthesis balancing program

    B. Syntax analyzer in compiler

    C. Keeping track of local variables at run time

    D. All of the above

    Answer: (D)

20. Consider the following operation performed on a stack of size 5.

```
Push(1);
Pop();
Push(2);
Push(3);
Pop();
Push(4);
```

```
Pop();
Pop();
Push(5);
```

After the completion of all operations, the number of elements present on the stack are

A. 1

B. 2

C. 3

D. 4

Answer: (A)

21. What is the name where deletion can be done in a linear list of elements from one end (front) and insertion can take place only at the other end (rear)?

 A. Queue

 B. Stack

 C. Tree

 D. Linked list

Answer: (A)

22. Let the following circular queue can accommodate maximum six elements with the following data

```
front = 2 rear = 4
queue = _______; L, M, N, ___, ___
```

What will happen after ADD O operation takes place?

A. front = 2 rear = 5queue = ______; L, M, N, O, ___

B. front = 3 rear = 5queue = L, M, N, O, ___

C. front = 3 rear = 4queue = ______; L, M, N, O, ___

D. front = 2 rear = 4queue = L, M, N, O, ___

Answer: (A)

23. If the MAX_SIZE is the size of the array used in the implementation of the circular queue. How is rear manipulated while inserting an element in the queue?

 A. rear=(rear%1)+MAX_SIZE

 B. rear=rear%(MAX_SIZE+1)

C. **rear=(rear+1)%MAX_SIZE**
D. rear=rear+(1%MAX_SIZE

Answer: (C)

24. What is the name where elements can be inserted or deleted, in a non-primitive linear data structure, from both the ends but not in the middle?

A. Queue
B. Circular queue
C. **Dequeue**
D. Priority queue

Answer: (C)

25. Consider the following pseudo-code. Assume that IntQueue is an integer queue. What does the function fun do?

```
void fun(int n)
{
    IntQueue q = new IntQueue();
    q.enqueue(0);
    q.enqueue(1);
    for (int i = 0; i < n; i++)
    {
        int a = q.dequeue();
        int b = q.dequeue();
        q.enqueue(b);
        q.enqueue(a + b);
        ptint(a);
    }
}
```

A. Prints numbers from 0 to n-1
B. Prints numbers from n-1 to 0
C. Prints first n Fibonacci numbers
D. Prints first n Fibonacci numbers in reverse order.

Answer: (C)

26. Suppose a circular queue of capacity (n – 1) elements is implemented with an array of n elements. Assume that the insertion and deletion operations are carried out using REAR and FRONT as array index variables, respectively. Initially, REAR = FRONT = 0. The conditions to detect queue full and queue empty are

A. Full: (REAR+1) mod n == FRONT, empty: REAR == FRONT

B. Full: (REAR+1) mod n == FRONT, empty: (FRONT+1) mod n == REAR
C. Full: REAR == FRONT, empty: (REAR+1) mod n == FRONT
D. Full: (FRONT+1) mod n == REAR, empty: REAR == FRONT

Answer: (A)
Explanation:
Suppose we start filling the queue. Let the maxQueueSize that is the capacity of the Queue is 4. Therefore, the size of the array that is used to implement, this circular queue is 5, which is n.

In the beginning when the queue is empty, FRONT and REAR point to 0 index in the array. REAR represents insertion at the REAR index. FRONT represents deletion from the FRONT index.

```
enqueue("a"); REAR = (REAR+1)%5; ( FRONT = 0, REAR = 1)
enqueue("b"); REAR = (REAR+1)%5; ( FRONT = 0, REAR = 2)
enqueue("c"); REAR = (REAR+1)%5; ( FRONT = 0, REAR = 3)
enqueue("d"); REAR = (REAR+1)%5; ( FRONT = 0, REAR = 4)
```

Now, the queue size is 4, which is equal to the maxQueueSize. Hence, the overflow condition is reached.

Now, we can check for the conditions.

When Queue Full:

```
(REAR+1) %n = (4+1) %5 = 0
FRONT is also 0.
Hence (REAR + 1) %n is equal to the FRONT.
When Queue Empty:
REAR was equal to FRONT when empty (because in the starting,
before filling the queue FRONT = REAR = 0)
Hence Option A is correct.
```

27. Following is a C like pseudo-code of a function that takes a queue as an argument and uses a stack S to do the processing.

```
void fun(Queue *Q)
{
    Stack S;  // Say it creates an empty stack S

    // Run while Q is not empty
    while (!isEmpty(Q))
    {
        // deQueue an item from Q and push the dequeued item
to S
        push(&S, deQueue(Q));
    }

    // Run while Stack S is not empty
```

```
    while (!isEmpty(&S))
    {
      // Pop an item from S and enqueue the poppped item to Q
      enQueue(Q, pop(&S));
    }
}
```

What does the above function do in general?

A. Removes the last from Q

B. Keeps the Q the same as it was before the call

C. Makes Q empty

D. Reverses the Q

Answer: (D)

Explanation: The function takes a queue Q as an argument. It dequeues all items of Q and pushes them to a stack S. Then pops all items of S and enqueues the items back to Q. Since the stack is in LIFO order, all items of the queue are reversed.

28. A data structure in which elements can be inserted or deleted at/from both the ends but not in the middle is?

A. Queue

B. Circular queue

C. Dequeue

D. Priority queue

Answer: (C)

29. A circular queue is implemented using an array of size 10. The array index starts with 0, the front is 6, and the rear is 9. The insertion of the next element takes place in the array index.

A. 0

B. 7

C. 9

D. 10

Answer: (A)

30. Trees and graphs are which type of data structures?

A. Linear data structures

B. Nonlinear data structures

C. Primitive data structures

D. None of the above

Answer: (B)

31. Consider the following pseudo-code. Assume that IntQueue is an integer queue. What does the function fun do?

```
void fun (int n)
{
    IntQueue q = new IntQueue();
    q.enqueue(0);
    q.enqueue(1);
    for (int i = 0; i < n; i++)
    {
        int a = q.dequeue();
        int b = q.dequeue();
        q.enqueue(b);
        q.enqueue(a + b);
        print(a);
    }
}
```

A. Prints numbers from 0 to n-1

B. Prints numbers from n-1 to 0

C. Prints first n Fibonacci numbers

D. Prints first n Fibonacci numbers in reverse order.

Answer: (C)

3

Void Pointer and Dynamic Memory Management

3.1 Void Pointer

The integer-type pointer points to value type integer variables, the float type pointer points to float value type variables and so on. In C, there is a general-purpose pointer that can point to any data type and is known as a void pointer. The void pointer is also called as generic pointer. A void pointer is one that does not have an associated data type. A void pointer can hold the address of any data type and can be typecast into any other data type.

Sometimes, we know we want a pointer, but we don't necessarily know what kind of data it points to. The C/C++ language allows a peculiar pointer called as void pointer that allows us to produce a pointer that is not type specific, meaning that it can be constrained to point to anything.

3.1.1 Why Void Pointer Is Useful?

malloc () is dynamic memory management function having return type as void *. malloc () function reserves the memory dynamically on heap partition and returns the base address of that memory location and we store that address in any pointer type variable. Here also we are not known of data type where we store our address, that is, in which type of data type is not known in prior. So, deciding which type of return type we assign to malloc () is very hard to decide so for that function return type must void * that is generic pointer is only the solution.

void pointer in C is used to implement generic functions in C language.

Syntax:

```
void * vptr;
```

DOI: 10.1201/9781003105800-3

A suitable typecast is must prior to dereferencing a pointer, as given below:

```
*((type  *)vptr);
↑    ↑__Typecast
Dereferencing operator
```

3.1.2 Program: Assign Float or Int Type Variables to Void Pointer and Typecasting When Dereferencing of Void Pointer to Its Particular Data Type

```
#include<stdio.h>
int main ()
{
int x = 100;
float y = 50.50;
int *iptr;
float *fptr;
void *vptr;
iptr = &x;
fptr = &y;
// iptr = &y;      /* Syntax error int pointer pointing to
float */
vptr = &x;         /* Valid, void pointer pointing to int  */
printf("\n  x= %d  " , *((int*)vptr) );     /* Typecast by int
type */
vptr = &y;                                  /* Valid , void
pointer pointing to float  */
printf("y= %f", *((float *)vptr));
return 0;
}
```

Output:

x=100
y=50.50

While printing value in void pointer, typecasting is must otherwise value of void pointer is not displayed.

Dereferencing a void pointer:

We use the indirection operator * to serve the purpose of dereferencing. However, in the case of a void pointer, we need to typecast the pointer variable to unreference it. This is because a void pointer does not have an associated data type. There is no way for the compiler to know or guess, which kind of data is pointed by the void pointer. Therefore, to take the data pointed to by a void pointer, we typecast it into the appropriate type of the data held in the void pointer's position.

3.1.3 Program: Address Void Pointer Assigned to Float or Int Pointer Type Variable and Dereferencing of Void Pointer

```
#include<stdio.h>
#include<conio.h>

//main function definition
void main()
{
int a=21;
float b=21.21;
void *ptr;  // Declaring a void pointer
ptr=&a;  // Assigning address of integer to void pointer.
printf("The value of integer variable a = %d",*( (int*) ptr)
);
// (int*)ptr - is used for typecasting. Whereas *((int*)ptr)
dereferences the typecasted void
// pointer variable.
ptr = &b;                                  // Assigning address
of float variable to void pointer.
printf("The value of float variable b= %f",*( (float*) ptr) );
}
```

Output:

The value of integer variable a=21

The value of float variable b=21.21

A void pointer may be useful if the programmer is not sure which type of data the end user has entered. In such a case, the programmer may use a void pointer to indicate the location of the unknown data type. The program can be set in such a way to ask the user to inform the type of data and typecasting can be carried out according to the information entered by the user. The following is a code snippet.

```
void funct(void *a, int z)
{
if(z==1)
printf("%d",*(int*)a);      // If user inputs 1, then it means
the data is an integer.
else if(z==2)
printf("%c",*(char*)a);   // Typecasting for character pointer.
else if(z==3)
printf("%f",*(float*)a);   // Typecasting for float pointer}
```

Another important point you need to keep in mind about void pointers is that pointer arithmetic cannot be performed on a void pointer.

Example:

```
void *ptr;
int a;
```

```
ptr=&a;
ptr++;
```

This statement is invalid and will result in an error because 'ptr' is a void pointer variable, and we cannot increment it.

3.1.4 Important Point

3.1.4.1. NULL versus uninitialized pointer: An uninitialized pointer stores an undefined value. A null pointer stores a defined value that is zero or NULL address.

3.1.4.2. NULL versus void pointer: Null pointer is a value that stores zero or NULL address, while the void pointer is a data type.

3.2 Pointer and Structure Data Type

3.2.1 Program: Implement a Program That Can Access Structure's Value Type Data Members by Using Structure's Pointer Type of Variable

```
#include <stdio.h>
struct student
{
char name [10];
int roll_no;
} stud;
int main ()
{
struct student *pt;
pt=&stud;
printf("Enter roll number \n");
scanf("%d",&stud.roll_no);        //or scanf("%d",&
pt->roll_no);
printf("Enter  name \n");
scanf("%s", stud.name);           //or scanf("%s",pt->name);
printf("\n Accessing student structure data members using
pointer variable ");
printf("\n Name: %s Roll Number: %d ", pt→name, pt→roll_no);
return 0;
}
```

Output:

Enter roll number
21

Enter name
xyz
Accessing student structure data members using pointer variable
Name: xyz Roll Number: 21

Note: -> (arrow operator) is used to access the structure's data member, using the structure's pointer type variable.

Instead of **pt->name** we can also use **(*pt).name**.

Here parentheses around *pt are necessary because the dot operator '.' has a higher precedence than '*'. If we write *pt.name, it will raise the syntactical error.

3.2.2 Program: Implement a Program That Can Access Structure's Pointer Type of Data Members by Using Structure's Pointer Type of Variable

```
#include <stdio.h>
struct student
{
char name[10];
int * roll_no;     //structure's pointer type data member
} stud;
int main()
{
int a=21;
struct student *pt;       // structure's pointer type variable
pt=&stud;
pt->roll_no =&a;
printf("Enter roll number \n");
scanf("%d",&(*(pt->roll_no)));
printf("Enter  name \n");
scanf("%s", pt->name);
printf("\n Accessing student structure data members using
structure's pointer variable ");
printf("\n Name: %s Roll Number: %d ", pt->name,
*(pt->roll_no));
return 0;
}
```

Output:

Enter roll number
21
Enter name
xyz
Accessing student structure data members using structure's pointer variable
Name: xyz Roll Number: 21.

3.2.3 Program: Structure's Pointer Variable Incrementation

```
#include <stdio.h>
struct student
{
char name[10];
int  roll_no;
} stud[2]={{"amol",21},{"suraj",31}};
int main()
{
struct student *pt;
pt=&stud[0];
printf("\n Accessing student structure roll number using
++pt->roll_no: ");
++pt->roll_no;
/* first pointer pt fetch the value of roll number and then
increment the roll number so output is 22 */
printf("\n Roll Number = %d",pt->roll_no);
printf("\n Accessing student structure roll number using
(++pt)->roll_no: ");
(++pt)->roll_no;
/* first address stored in pointer pt increment and then shows
value of roll number so output is 31 */
printf("\n Roll Number = %d",pt->roll_no);
return 0;
}
```

Output:

Accessing student structure roll number using ++pt->roll_no:
Roll Number=22.
Accessing student structure roll number using (++pt)->roll_no:
Roll Number=31.

Explanation:

++pt→roll_no;

Here first calculate the pt->roll_no value and then that roll number value gets incremented. Therefore, pt->roll_no value is 21 then incremented by 1 so it becomes 22.

However (++pt) →roll_no;

Increment pt first means increment address by 1 means one location and then gives the count structure data member's value at that address location which is 31.

3.3 Dynamic Memory Allocation

In C language, number of elements that are size of the array to be specified at compile time.

The initial judgement of size, if it is erroneous, can cause the failure of a program or the waste of memory space. The process of memory allocation on execution is called dynamic memory allocation. 'C' provides **"memory management function"** that can be used for allocating and freeing memory during program execution. For dynamic memory management, **Heap memory partition of RAM is used**.

Memory allocation function:

Function	Task
malloc	The malloc () function allocates requested size in bytes for single element and returns a pointer to first byte of allocated space and **initial element value is garbage**.
calloc	The calloc () function assigns space for an array of elements, initializes them to zero and returns a pointer to the first byte of allocated space.
free	The free () function releases the space previously assigned by the malloc (), calloc () or realloc () function.
realloc	The realloc () function alters the size of the space already assigned.

3.3.1 Allocating a Block of Memory: malloc () Function

General syntax or function prototype of malloc () function:

```
ptr= (cast_type *) malloc(byte_size);
```

The malloc () function, returns the void * so if catching variable is int * then typecasting is must.

Example:

```
char *cptr;
cptr=(char*) malloc (10);
```

Allocates 10 bytes of space for characters and store address of first byte in the pointer cptr pointer variable, this is illustrated as:

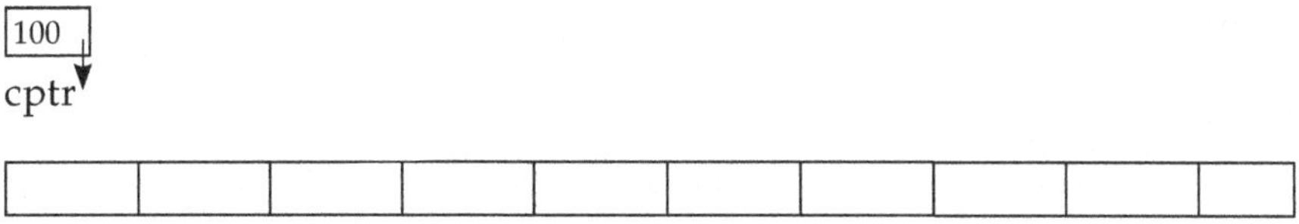

100

cptr

100 10 byte of space on heap partition
109

The dynamically assigned storage space does not have a name, and therefore its contents can only be accessed by a pointer.

Structure_pointer_variable=(struct stud *) malloc (sizeof(struct stud)));
where Structure_pointer_variable is a **pointer of type struct stud**.

Remember, **malloc () function** allocates a block of **contiguous bytes.**

Allocation can fail if space on heap is not sufficient to satisfy the request. If it fails, it returns a **NULL address.**

3.3.2 Allocation of Multiple Block of Memory: calloc () Function

The calloc () function allocates a multiple block of storage, each of the same size, and then set all blocks of bytes to zero.

```
struct  stud *s;
s=(struct stud*)calloc(30,sizeof(struct
stud) );
```

General syntax of calloc function:
ptr =(cast_type *) calloc (n, element_size);
where n=Number of elements
element_size=Size of each element.

3.3.3 Releasing the Used Space Using free () Function

Compile time storage of a variable is allocated and released by the compiler depends on the scope of variable and lifetime of variable. With dynamic or run time memory allocation, the programmer is responsible for releasing the space when it is not required.

General syntax or function prototype of free function:
void free(void *ptr);

The free () function does not release memory that is allocated to that pointer variable but released the memory pointed by that pointer variable on the heap.

The pointer variable is either local, global or static, so it is placed either in local memory, that is stack memory, global or in static memory, respectively. Memory allocated to the pointer variable is deallocated depending on its lifetime or aliveness. The malloc () function allocates memory of memory.

Program: The following program shows the use of malloc () and free () function.

```
#include <stdio.h>
#include <stdlib.h>
#include<string.h>

// main function definition
int main()
{
   char *str;

   /* Initial memory allocation using malloc function */
   str = (char *) malloc (15);
```

```
    strcpy(str, "Amol Jagtap");
    printf("String = %s,  Address = %u\n", str, str);
    free(str);    // deallocate memory from heap partition which
is reversed by malloc function
    return 0;
}
```

Let us compile and run the above program, this will produce the following result:

Output:

String=Amol Jagtap, Address=355090448

3.3.4 The realloc () Function

The realloc () function alters the size of a memory block that was previously assigned with malloc () or calloc ().

General syntax or function prototype of realloc function:

void * realloc (void *ptr, size_t size);

The ptr argument serves as a pointer to the original memory block. The new size, in bytes, is specified by size. There are several possible outcomes with realloc () function. If there is enough space to expand the memory block pointed to by ptr, the additional memory is allocated and the function returns ptr.

If adequate space does not exist to expand the current block in its current location, a new block of the size is allocated, and existing data are copied from the old block to the beginning of the new block. The old block is released, and the function returns a pointer to the new block.

If the ptr argument is NULL, the function acts like malloc (), allocating a block of size bytes and returning a pointer to it. If the argument size is 0, the memory that ptr points to is freed, and the function returns NULL.

If memory is not sufficient for reallocation either by expanding the old block or by allocating a new one, the function returns NULL.

Program: Implement a program using realloc() to increase the size of a block of dynamically allocated memory.

```
#include <stdio.h>
#include <stdlib.h>
#include <string.h>

//main function definition
int main ()
{
char buf[80], *message;
/* Enter any string */
puts("Enter a line of text.");
gets(buf);
/* Allocate the initial block and copy the string to it
typecast char * is must */
```

```
message = (char *) realloc(NULL, strlen(buf)+1);
strcpy(message, buf);
/* Display the entered message on console */
puts(message);
/* Take another string from the user. */
puts("Enter another line of text.");
gets(buf);
/* Increase the space for the string dynamically, then
concatenate the string to it. */
message = (char *) realloc(message,(strlen(message) +
strlen(buf)+1));
strcat(message, buf);
/* Display the concatenated message on the console */
puts(message);
return 0;
 }
```

Output:

Enter a line of text.
Hello students.
Hello students.
Enter another line of text.
Welcome to our class.
Hello students. Welcome to our class.

3.4 Memory Leakage

A memory leak occurs when a programmer creates a memory dynamically on heap partition and forget to delete it. Here memory allocated at the time of compilation is removed by the runtime environment based on the variable's lifetime. However, memory allocated dynamically is not deleted automatically, for that purpose programmer must write code for that, in C language free () function is used for releasing dynamically allocated memory.

3.4.1 Program: To Demonstrate a Function with Memory Leak

```
#include <stdlib.h>
#include<stdio.h>
void fun()
{
int *ptr = (int *) malloc(sizeof(int));
printf("\n Memory allocated dynamically for integer
variable");
}
int main()
```

```
{
fun();
return 0;
}
```

Here, in the function, fun, memory allocated dynamically for int but which is not releasing so that memory on heap remains allocated after the total execution of the program. For that purpose, free function is used. To avoid memory leaks, memory allocated on the heap must always be freed when no longer needed.

3.4.2 Program: To Demonstrate a Function without Memory Leak

```
#include <stdlib.h>
#include<stdio.h>
void fun()
{
int *ptr = (int *) malloc(sizeof(int));
printf("\n Memory allocated dynamically for integer
variable");
free (ptr);   // release dynamically allocated memory for
integer variable
}
int main()
{
fun();
return 0;
}
```

Using free function, we have to release the dynamically allocated memory on heap. Therefore, this memory on the heap partition is again used by any other variable and the memory leakage problem is solved.

3.5 Dangling Pointer

A dangling pointer points to a memory location that has previously been freed. Storage is not available anymore. Trying to access that memory can cause a segmentation error or can crash the program.

The pointer can act as a dangling pointer in three ways, which are explained below:

3.5.1 Deallocating a Memory Pointed by Pointer Causes Dangling Pointer

```
#include <stdlib.h>
```

```
#include <stdio.h>
int main()
{
int *ptr = (int *)malloc(sizeof(int));
/* After free function call, pointer ptr becomes a dangling
pointer because dynamically allocated memory to ptr becomes
free and ptr points to that memory location on heap which is
not allocated. */
free(ptr);
// Assign NULL address to ptr pointer, now ptr pointer is not
a dangling pointer.
ptr = NULL;
return 0;
}
```

3.5.2 Function Call

The pointer pointing to local variable becomes dangling when a local variable is non-static and which returns the address to another function.

```
#include <stdio.h>
int * sample()
{
/* x is a local variable and goes out of scope and variable
becomes dead after execution of sample() function is over. */
int x = 21;
return &x;
}
int main ()
{
int *p = sample ();
fflush(stdin);
// p points to something which is not valid anymore.
printf("%d", *p);
return 0;
}
```

Output:
Segmentation Fault

As runtime error.

The solution to this problem:

The pointer pointing to local variable does not become dangling when a local variable is static.

```
#include <stdio.h>
int * sample()
{
```

```
/* x is local variable but it is static so variable does not
become dead after an execution of sample() function is over.
Value and memory reserved for variable x remains live
throughout program execution */
static int x = 21;
return &x;
}
int main()
{
int *p = sample();
fflush(stdin);
// Not a dangling pointer as it points to static variable.
printf("%d", *p);
return 0;
}
```

Output:
21

3.5.3 When Local Variable Goes Out of Scope or Lifetime Is Over

```
#include <stdio.h>
int main()
{
int *ptr;
{
int i;
ptr = &i;     //Block1
}
return 0;
}
```

Here ptr pointer is dangling pointer because after block1 variable i value becomes dead and also variable i's scope is over. That is again pointer ptr stores address of freed variable. Therefore, ptr becomes dangling after block1's execution is over.

3.6 Interview Questions

1. What is the use of void pointer?
2. Explain the generic pointer with a suitable example.
3. Differentiate between the void pointer, NULL pointer and an uninitialized pointer.

4. Write a note on dynamic memory allocation with a suitable example.
5. What are the different dynamic memory allocation functions and explain with suitable example?
6. Write a note on memory leakage.
7. What is the use malloc (), calloc (), realloc () and free () functions?
8. What are the differences between dangling pointer and memory leak?
9. What is mean by dangling pointer explain with an example?
10. What is the dangling pointer and how to avoid it?
11. What are the different proposed solutions to avoid the dangling pointers?
12. How to avoid dangling pointer issues and memory leakage problem?
13. What is the difference between malloc() and calloc() dynamic memory management functions?

3.7 Multiple Choice Questions

1. What is the return type of malloc() function?

   A. float

   B. int *

   C. void *

   D. double *

   Answer: (C)
   Explanation:
   malloc() function having return type as void *.

2. Which of the below header files is required to be included to use dynamic memory allocation functions?

   A. stdlib.h

   B. stdio.h

   C. memory.h

   D. dos.h

   Answer: (A)
   Explanation:
   stdlib.h is a header file which consists of the declaration for dynamic memory allocation functions such as malloc(), calloc(), realloc() and free() function.

3. Which memory partition stores dynamically allocated variables?
   A. **Heap memory partition**
   B. Stack memory partition
   C. Global memory partition
   D. Code memory partition

   Answer: (A)
   Explanation:
   Heap memory partition is used for dynamic memory management.
4. Which method is used to remove the dynamically allocated memory space?
   A. dealloc() function
   B. **free() function**
   C. realloc() function
   D. Both A and B

   Answer: (B)
   Explanation:
   free() method is used to release the memory space allocated on heap partition by malloc() and calloc() functions.
5. Which of the following is correct function declaration for malloc() function in C language?
   A. **void * malloc (int size_t)**
   B. char* malloc (char)
   C. int * malloc (int size_t)
   D. float * malloc (float size_t)

   Answer: (A)
   Explanation:
   malloc() function having return type as void * and argument to the malloc() function is size in integer.
6. Which functions are used for memory allocation at runtime in C language?
   A. malloc() function
   B. calloc ( ) function
   C. realloc () function
   D. **Both A and B**

   Answer: (D)
   Explanation:

Dynamic memory allocation is done through malloc() and calloc() library functions.

7. What is the problem with following code?

```
#include <stdlib.h>
#include<stdio.h>
void fun()
{
int *ptr = (int *) malloc(sizeof(int));
printf("\n Memory allocated dynamically for integer
variable");
}
int main()
{
fun();
return 0;
}
```

A. Memory Leakage

B. Dangling pointer

C. Compiler error

D. None of the the above

Answer: (A)

Explanation:

The problem is memory leakage, pointer ptr is allocated some memory which is not freed meaning that what remains allocated on the heap memory partition causes memory leakage.

8. Consider the following program, where are x, y and z are stored in main memory partition?

```
int y;
int main()
{
int z = 0;
int *x = int * malloc( sizeof(int));
}
```

A. x is stored on the heap memory partition, y is stored on the global partition and z is stored on stack partition.

B. x, y and z variables are stored on the stack partition.

C. x is stored on the heap memory, y and z variables are stored on stack partition.

D. x and y are stored on the heap memory partition and z is stored on stack partition.

Answer: (A)
Explanation:
For x variable. dynamic memory gets reserved so it is allocated on the heap memory partition. While y is global variable and stored on global partition of the RAM. z is a local variable and stored on stack partition of the RAM.

9. Which of the following functions allocates multiple blocks of memory at runtime, each block allocated having same size and each byte of allocated space is initialized to zero?
   A. malloc()
   B. realloc()
   C. calloc()
   D. free()

   Answer: (C)
   Explanation:
   calloc() library function allocates multiple blocks of memory at runtime, each block allocated having same size and each byte of allocated space is initialized to zero.

10. Memory is reserved dynamically using malloc() or calloc() functions but memory area that is allocated dynamically is not accessible to programmer any more afterward is called as?
    A. Dangling memory
    B. Dangling pointer
    C. Memory leakage
    D. Pointer leakage

    Answer: (C)
    Explanation:
    Memory leak occurs when programmer creates a memory dynamically using malloc() and calloc() functions on heap partition and forget to delete it or NULL address is assigned to that pointer then reserved memory on heap remains allocated and cannot be used by anyone this problem is called as memory leakage.

11. In realloc() library function, if the new size of the memory block is greater than the old size, then newly added memory space .....?
    A. is initialized to one
    B. is initialized to zero
    C. is not initialized
    D. none of the above

Answer: (C)
Explanation:
In realloc() library function, newly added space is not initialized.

12. What is dangling pointer?
    A. if pointer is assigned NULL value
    B. if pointer points to a memory location that has already been freed
    C. if pointer is assigned to more than one variable
    D. None of the above

    Answer: (B)
    Explanation:
    If pointer is **pointing** to a memory location from where a variable has been already deleted.

13. The free() function frees the memory allocated at runtime and returns _____
    A. pointer value
    B. the memory address of the freed variable
    C. no value is return
    D. an float value is return

    Answer: (C)
    Explanation:
    Free function having return type as void means it does not return any value.

14. Which of the following function allocates requested size of bytes for single element and returns a pointer to first byte of allocated space at runtime and initial element value is garbage.
    A. calloc() function
    B. malloc() function
    C. realloc() function
    D. None of the above

    Answer: (B)
    Explanation:
    malloc() function allocates memory dynamically and returns an address of first byte and initial element value is garbage.

15. Which of the following pointer stores a defined value that is zero or NULL address.

    A. NULL pointer

    B. void pointer

    C. Dangling pointer

    D. None of the above

    Answer: (A)
    Explanation:
    Null pointer is a value that stores zero or NULL address.

16. Which of the following statements are correct?

    Statement 1: A NULL pointer stores a defined value that is zero or NULL address.

    Statement 2: The void pointer is a data type while NULL pointer is not a data type.

    A. Statement 1 is false and Statement 2 is true

    B. Statement 2 is false and Statement 1 is true

    C. Statements 1 and 2 are false

    D. Statements 1 and 2 are true

    Answer: (D)
    Explanation:
    The void pointer is a data type and NULL pointer stores NULL address or zero.

17. Which of the following statements are true or false?

    Statement 1: malloc () is static memory management function having return type as void *.

    Statement 2: malloc () function reserves the memory dynamically on heap partition.

    A. Statement 1 is false and Statement 2 is true

    B. Statement 2 is false and Statement 1 is true

    C. Statements 1 and 2 are false

    D. Statements 1 and 2 are true

    Answer: (A)
    Explanation:
    malloc () is dynamic memory management function having return type as void * and reserves the memory on heap partition.

18. Which of the following statements are true or false?

    Statement 1: The realloc () function alters the size of the space already assigned.

    Statement 2: The malloc () function assigns space for an array of elements, initializes them to zero.

    A. Statement 1 is false and Statement 2 is true
    B. **Statement 2 is false and Statement 1 is true**
    C. Statements 1 and 2 are false
    D. Statements 1 and 2 are true

    Answer: (B)
    Explanation:
    The calloc () function assigns space for an array of elements, initializes them to zero.

19. Which of the following statements are true or false?

    Statement 1: Compile time storage of a variable is allocated and released by the compiler depends on the scope of variable and lifetime of a variable.

    Statement 2: In dynamic allocation, the programmer is responsible to release the space when it is not required.

    A. Statement 1 is false and Statement 2 is true
    B. Statement 2 is false and Statement 1 is true
    C. Statements 1 and 2 are false
    D. **Statements 1 and 2 are true**

    Answer: (D)
    Explanation:
    In dynamic allocation, the programmer is responsible to release the space and static memory is released by compiler depends on the scope and lifetime of a variable.

20. Which of the following statements are true or false?

    Statement 1: A memory leak occurs when a programmer creates a memory dynamically on heap partition and forget to delete it.

    Statement 2: A dangling pointer points to a memory location that has previously been freed.

    A. Statement 1 is false and Statement 2 is true
    B. Statement 2 is false and Statement 1 is true
    C. Statements 1 and 2 are false
    D. **Statements 1 and 2 are true**

Answer: (D)
Explanation:
A dangling pointer points to a memory location that has previously been freed and a memory leak occurs when a programmer creates a memory dynamically on heap partition and forget to delete it.

21. Which of the following statements are true or false?

Statement 1: The free () function releases the space previously assigned by the malloc (), calloc () or realloc () function.

Statement 2: A dangling pointer and void pointers both are same.

A. Statement 1 is false and Statement 2 is true

B. Statement 2 is false and Statement 1 is true

C. Statements 1 and 2 are false

D. Statements 1 and 2 are true

Answer: (D)
Explanation:

The concepts of a dangling pointer and void pointers are different.

4

Linked Representation of Linear Data Structures

4.1 Limitations of Static Memory Allocation and Advantages of Dynamic Memory Management

4.1.1 Introduction

In Chapter 3, we have studied arrays and static memory allocation. We have learned that elements are stored sequentially in an array, and we can have sequential as well as random access to the elements of arrays and static memory allocation for the array elements. All of the memory management we have studied so far has been static memory management, which means that memory is allocated at the compile time rather than at run time. Memory in array is reserved at compile time, which is known as static memory management. While in a linked list, memory is reserved at run time, which is known as dynamic memory management.

4.1.2 Major Limitations of Static Memory Allocation

If more static data space or static memory is declared than is needed, there is a waste of memory space. If there is less static space or static memory is declared than is needed, then it becomes impossible to expand this fixed size during run time. In an array, inserting and deleting elements are relatively expensive.

4.1.3 Major Advantages of Dynamic Memory Allocation

Using dynamic memory management, data structures can grow and shrink to fit changing data requirements. We can allocate additional storage whenever we need them at run time. We can de-allocate or free space dynamically whenever we use space and now that space is no longer required.

DOI: 10.1201/9781003105800-4

However, the major drawback of array as the data structure is as follows:

Once elements are stored sequentially, it becomes very difficult to insert an element in between or to delete the middle element. For the insertion of new elements, elements should be shifted in an upward direction. If we delete any element from the array, a vacant space will be created in the array. Shifting elements will be **time- and logic-**consuming, and it is **time and space inefficient**.

In an array data structure, we cannot make an exact judgment about the array size. Sometimes, memory may be wasted or there may be a case in which we have to increase the array size.

Therefore, to solve the above problems, we will learn a new data structure that is a linked list.

With a linked list, insertions and deletions can be efficiently carried out. Linked lists are dynamic data structures, which can expand or shrink while a program is running. Efficient memory utilization is done using dynamic memory management. Many complicated applications can be easily performed with linked lists.

4.2 Concept of the Linked List

4.2.1 Introduction

A linked list is a set of nodes where each node has **two fields 'data' and 'link' or 'next'.**

The 'data' field stores actual piece of information and the 'link' or 'next' field is used to point to the next node or store the address of the next node (Figure 4.1).

Note that the link field of the last node consists of 'NULL', which indicates the end of the singly linear linked list. NULL denotes Null pointer or Null address. Null pointer means last node of linked list does not contain addresses of the next node.

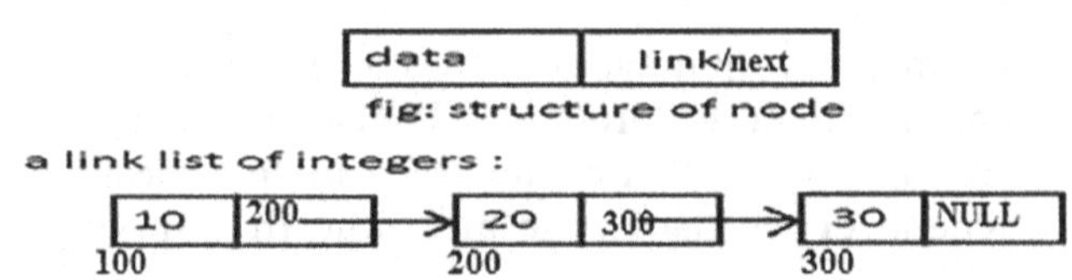

FIGURE 4.1
Structure of node and linked list of integers.

4.2.2 Singly Linked List Representation Using C Language

If we need to store integer numbers as data in the linked list, then the linked list representation of the memory has the following declaration syntax.

```
struct node
{
      int data;
      struct node *next;
};
```

Using the above declaration syntax, we can create a number of nodes having integer data and the next pointer stores the address of the next node in the linked list.

4.2.3 Operations on the Linked List

The elementary operations performed on the linked list are given below:

1. Creation 2. Insertion 3. Deletion 4. Traversing 5. Searching 6. Display

Let us define these above operations in detail as,

1. **Creation:** This operation is used to create a new linked list.
2. **Insertion:** This operation is used to insert a new node at the specified position, either at the start, end or at the specified location.
3. **Deletion:** This operation is used to delete a node from the start, end or specified location.
4. **Traversing:** This operation is used to visit all nodes in the linked list from either the start or the end of the list.
5. **Searching:** This operation is used to find the specific node data is present in the linked list or not.
6. **Display:** This operation is used to print all node data in the linked list from the start or end of the node.

4.2.4 Program: Implementation of Static Linked List Containing Three Nodes and Displaying Values on Those Nodes

```
#include<stdio.h>
#include<conio.h>
 struct node  //structure for a node
{
      int data;
```

```
        struct node *next;
};
struct node n1,n2,n3;  //three structure variables for
creating linked list nodes
struct node *first,*temp; //structure's pointer type variables
to store address of first node          //and temporary node in
the linked list
int main()
{
        n1.data=10; //insert 10 data into the first node n1
        n1.next=&n2;  //attach node second that is n2 to the
first node n1 by coping address  //of n2 into the next field
of node n1
        n2.data=20;
        n2.next=&n3;
        n3.data=30;
        n3.next=NULL;  //copy address 'NULL' into the n3 node's
next field
        first=&n1;      // assign address of n1 node into the
structure's start pointer
        temp= first;   //assign address of first node into the
temp pointer for traversing
        while(temp!=NULL)  //till temp pointer does not contains
address NULL executes the            //while loop
        {
printf("\n%d",temp->data); //display data into the temp node
temp=temp->next;  //move temp pointer to the next node into
the linked list
        }
        return 0;
}
```

Output:

10
20
30

Explanation:

Step 1: Declare the nodes n1, n2 and n3 in the structure type (Figure 4.2).

Step 2: Start filling data into each node in the data field and assigning the address of the next node to the next pointer field of the node.

Here, node n2's address is 200 and is assigned to the next field of node n1, which is shown in Figure 4.3.

data	next

FIGURE 4.2
Structure of node in a linked list.

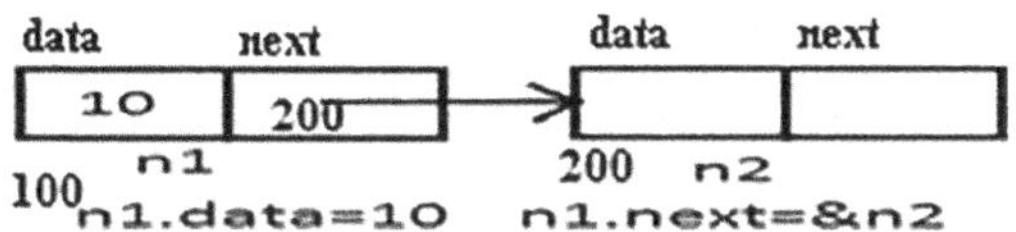

FIGURE 4.3
Insertion of data and address in a node.

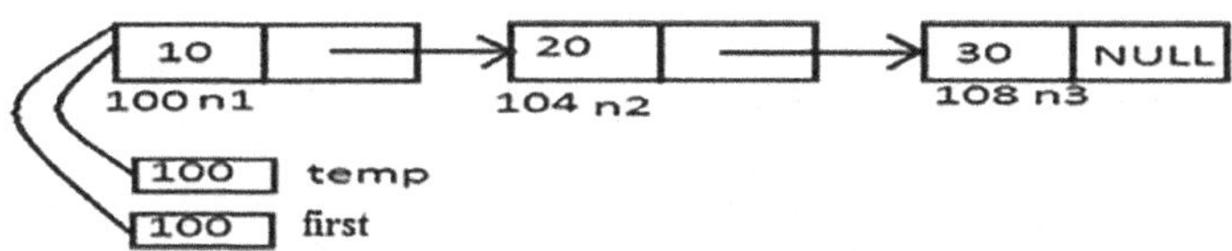

FIGURE 4.4
Linked list contains three nodes with data and address of next node.

Similarly, add 20 data in the data field and address of the n3 node to the n2 node's next field and 30 data to the n3 node and the last node's link or next field contains a NULL address assigned using the statement, n3. next=NULL;

Step 3: Set

```
first=&n1;
temp=first;
```

Step 4: while (temp!=NULL)

Traverse the linked list up to last node and print the data field in each node as:

```
{
        printf("%d", temp->data);
        temp=temp->next;
}
```

temp is the pointer of node type so it stores the address of node.

```
temp = temp ->next;
```

Here, **temp ->next** is the address of the next node in the linked list.

When the temp pointer points to the last node, then temp->next value is NULL which is assigned to temp by using above statement

```
temp=temp->next ;
And then while (temp != NULL)
```

This condition becomes false and the loop gets terminated.

4.3 Types of Linked List

1. Singly linear linked list
2. Singly circular linked list
3. Doubly linear linked list
4. Doubly circular linked list

4.3.1 Singly Linear Linked List

- Singly because node consists of only one link part that is next which stores address of the next node.
- Linear because the last node points to nothing means last node's next or link field contains a NULL address.
- The first node is called the head or the first node (Figure 4.5).

4.3.2 Singly Circular Linked List

- Circular means the last node's link or next field points to the first or head node or contains the address of the first node.
- That is last node n3's next field contains the address 100 of first node containing data 10 (Figure 4.6).

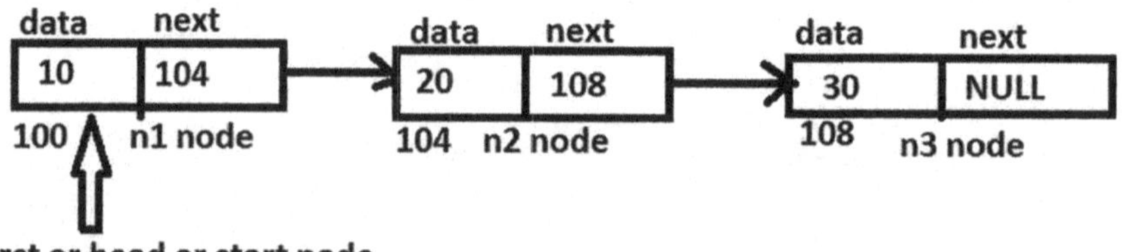

FIGURE 4.5
Singly linear linked list.

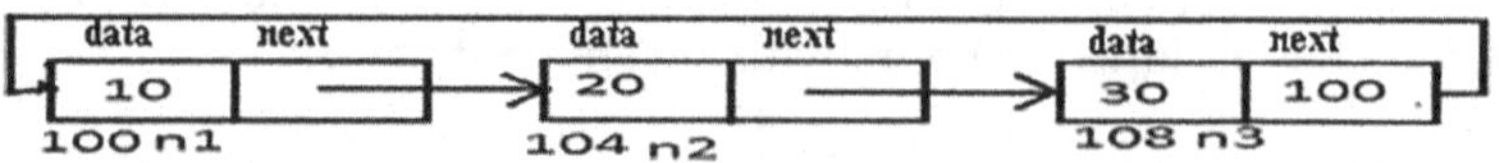

FIGURE 4.6
Singly circular linked list.

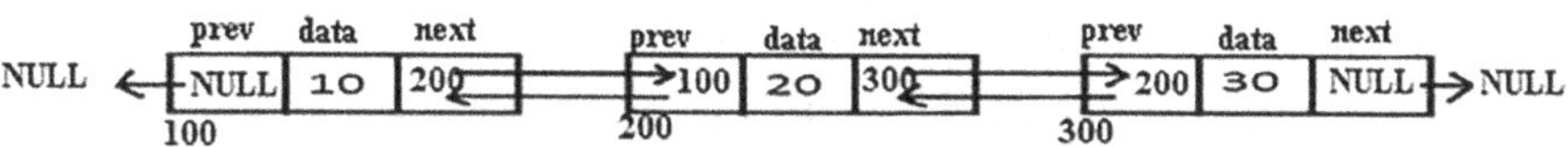

FIGURE 4.7
Doubly linear linked list.

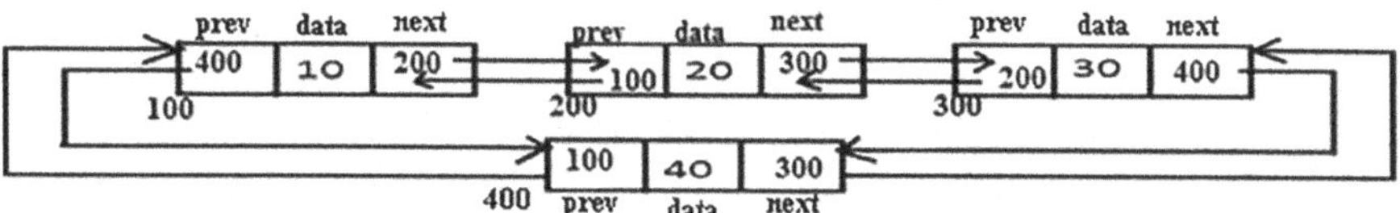

FIGURE 4.8
Doubly circular linked list.

4.3.3 Doubly Linear Linked List

- Doubly because each node has two pointers previous and next pointers.
- The previous pointer points to the previous node and the next pointer points to the next node or contains the address of next node.
- Only the head node's previous pointer field contains NULL and the last node's next pointer points to a NULL address (Figure 4.7).

4.3.4 Doubly Circular Linked list

- **In a circular doubly linked list, the next pointer of the last node points to the previous pointer of head and the previous pointer of the head node points to the next pointer of the last node.**
- The head node is a special node that may have either dummy data or some useful information, such as the total number of nodes in the list (Figure 4.8).

4.4 Singly Linear Linked List

Singly linear linked list consists of only one link field that is next, which stores the address of the next node and the last node's next or link field contains a NULL address.

4.4.1 Algorithm for Singly Linear Linked List

4.4.1.1 Algorithm: Insert a Node at the Last of Singly Linear Linked List

- Step 1: [Check for overflow]

```
If new1 = NULL, then
            Print, "over flow"
            Exit
Else
            New1 = (stuct node *)malloc (sizeof(struct
            node));
End if
```

- Step 2: Set new1 -> data=num
- Step 3: Set new1 -> next=NULL
- Step 4: If start == NULL, then set start=new1 i.e. assign address of newly created node to start pointer.
- Step 5: If start != NULL, then search the last node in the linked list using temp pointer until

```
temp->next != NULL .
```

- Step 6: Set temp->next=new1.
- Step 7: Stop.

4.4.1.2 Algorithm: Delete a Node from the Singly Linear Linked List

Step 1: [check for under flow]

```
If start == NULL, then
        Print "underflow"
Exit
```

Step 2: Read the information which you want to delete, say 'num'.
Go on checking **the data field of each node**, whether it matches 'num',

```
while((temp != NULL)&&(found == 0))
{
if(temp -> data == num)
      {
      found =1;
      }
      else
      {
```

```
      temp =temp->next;
      }
  }
```

Step 3: If found == 0
  then print "No such number in the linked list" .
  Step 4: If the node is found, then store that **found the node address into a temp pointer.** If the found node is the first node that is, temp == start, then simply move start one position ahead and remove temp node.

```
        start = start->next ;
        temp->next = NULL;
        free(temp);
```

Step 5: Else node found is other than first node, then another pointer called as prev is point to the previous node of temp node which you want to delete as

```
prev=start;
while(prev->next != temp)
{
        prev = prev->next;
}
```

Then

```
        prev->next=temp->next;
```

That is pointer prev's next part (prev->next) points to the 'temp' node's (found node's) next part & make temp->next=NULL (i.e. 'temp' node cannot point to any other node) then the memory allocated for that 'temp' node is deleted by free(temp) function.

4.4.1.3 Algorithm: Display Elements into Linked List

Step 1: If start==NULL then
  Linked list is empty.
Step 2: Otherwise, list is not empty, then move node pointer around list, visit each node display the information

```
        temp=start;
        while(temp!=NULL)
        {
                        printf("%d   ",temp->data);
                        temp=temp->next;
}
```

4.4.2 Program: Implementation of Singly Linear Linked List

```
//header files included
#include<stdio.h>
#include<stdlib.h>
//Function Prototype
void create_node();
void display();
void delete_node();
//structure for node in singly linear linked list
struct node
{
 int data;
 struct node *next;
}*new1,*start,*prev,*temp;
//new1 pointer stores address of newly created node
//start pointer stores address of start node in linked list
//prev pointer stores address of previous node which we want
to delete in linked list
//temp pointer is used for traversing purpose
int main()
{
 int ans;
 char ch;
 start=NULL;
 do
 {
  printf("Singly Linear Linked List: \n");
  printf("\t1.create_node\t2.diplay\t3.delete_node\t4.exit ");
  printf("\n Enter your choice: ");
  scanf("%d",&ans);
  switch(ans)
  {
      case 1:create_node();
      break;
      case 2:display();
      break;
      case 3:delete_node();
      break;
      case 4:exit(0);
      default: printf("\n Enter correct choice");
  }
  printf("Do you want to continue press[yes(y) /no(n)] : ");
  fflush(stdin);
  scanf("%c",&ch);
 } while(ch=='y'||ch=='Y');
 return 0;
}
//create a node and add new node at the last in the Singly
Linear Linked List
```

```
void create_node()
{
//allocate dynamic memory for a single node
 new1=(struct node*)malloc(sizeof(struct node));
 printf("Enter data : ");
 scanf("%d",&new1->data);
 new1->next=NULL;
      if(start==NULL)  //If Linked List is empty
      {
      start=new1;
      }
      else    //If Linked List having at least one node
      {
      temp=start;
      while(temp->next!=NULL)     //traverse the list  until
      last node reached
      {
      temp=temp->next;
      }
      temp->next=new1;
      }
printf("Node inserted at the end of Linked list=%d
\n",new1->data);
}
//display nodes in Singly Linear Linked List
void display()
{
 if(start==NULL)
 {
  printf("Linked list  is not created.\n");
 }
 else
 {
  temp=start;
printf("Data in the Singly Linear Linked List is as follows:
\n");
while(temp!=NULL)
{
      printf("%d \t",temp->data);
      temp=temp->next;
      }
      printf("\n");
  }
}
//delete a node from a Singly Linear Linked List
void delete_node()
{
 int num;
 int found=0;
 temp=start;
```

```
 if(start==NULL)    //Linked list is not created
 {
  printf("\n Linked list is not created ,first create a
node.\n");
 }
 else
 {
  printf("Enter value which you want to be deleted: ");
  scanf("%d", &num);
  while((temp!= NULL)&&(found == 0))  //loop for searching the
data which we want to      {
//delete and store address of deleted node into
//temp pointer.
      {
      if(temp -> data == num)
      {
      found =1;
      }
      else
      {
      temp =temp->next;
      }

  }
  if(found == 0)      // If node is not found
  {
      printf("\n Node is not present in the Linked List. \n");
  }
  else
  {         //If node is found then node which we want  to
delete is first      // node or other than first node.
      if(start == temp)    //If deleted node is First node in
      Linked List
      {
      start = start->next;
      temp->next=NULL;
      free(temp);
      temp=NULL;
      }
      else   //If deleted node is other than first node in
      Linked List
      {
      prev=start;
      while(prev->next!=temp)
      {
       prev=prev->next;
      }
      prev->next=temp->next;
      temp->next=NULL;
```

```
        free(temp);
        temp=NULL;
        }
        printf("Node deleted =%d\n",num);
    }
}
}
```

Output:

```
Singly Linear Linked List:
        1.create_node   2.diplay        3.delete_node   4.exit
 Enter your choice: 1
Enter data : 21
Node inserted at the end of Linked list=21
Do you want to continue press[yes(y) /no(n)] : y
Singly Linear Linked List:
        1.create_node   2.diplay        3.delete_node   4.exit
 Enter your choice: 1
Enter data : 31
Node inserted at the end of Linked list=31
Do you want to continue press[yes(y) /no(n)] : y
Singly Linear Linked List:
        1.create_node   2.diplay        3.delete_node   4.exit
 Enter your choice: 2
Data in the Singly Linear Linked List is as follows:
21      31
Do you want to continue press[yes(y) /no(n)] : y
Singly Linear Linked List:
        1.create_node   2.diplay        3.delete_node   4.exit
 Enter your choice: 3
Enter value which you want to be deleted: 31
Node deleted =31
Do you want to continue press[yes(y) /no(n)] : n_
```

4.5 Singly Circular Linked List

Singly circular linked list is a list in which last node's next pointer always point's to first or head node and node contains only one link or next field.

Operations on Linear Circular Linked List:

i. Creation
ii. Display
iii. Deletion

4.5.1 Algorithm for Singly Circular Linked List

4.5.1.1 Creation of Singly Circular Linked List

Step 1 (Figure 4.9):

Entering the data in data or info field of the node currently it is 10 and assign the NULL address to the next field in the node (Figure 4.10).

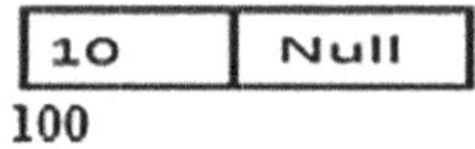

FIGURE 4.9
Create a node in the singly circular linked list.

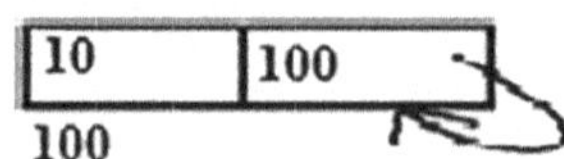

FIGURE 4.10
Insert first node in the singly circular linked list.

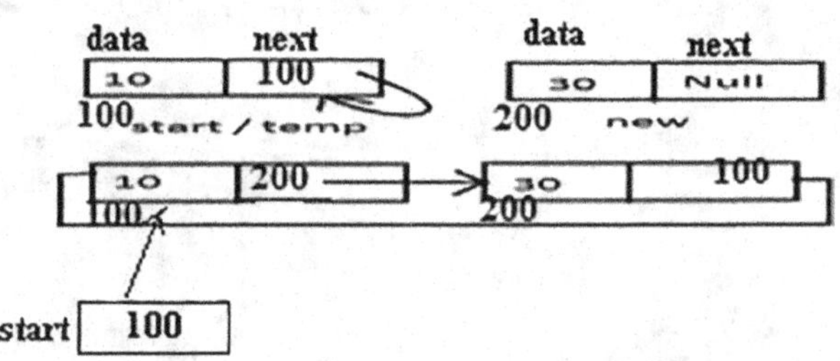

FIGURE 4.11
Insert second node in the singly circular linked list.

If flag=1 or start pointer contains a NULL address that means the newly created node is the first node in a circular linked list, then assign address of the new node to the pointer start and the new node's next contains address of first node means same node points to itself. Here new node is the first node in a circular linked list, then the first node's next field stores address of the first node that is 100 and set flag=0 means next time onwards we can insert new node at the end of the circular linked list.

Step 2: Create new node and insert data as 30 in the data field and insert new node in previously created circular linked list as shown in Figure 4.11.

Here, store the address of the new node into the new pointer of the circular linked list and the address of the start node is stored in the new node's next field. Thus, the new node is inserted into the circular lined list.

4.5.1.2 Display Elements in Circular Linked List

Step 1: The circular linked list contains three nodes as shown in Figure 4.12, which is given below:

Step 2: First store the address of the start node in temp pointer. Traverse the temp pointer till temp pointer contains the address of the last node by using the following code:

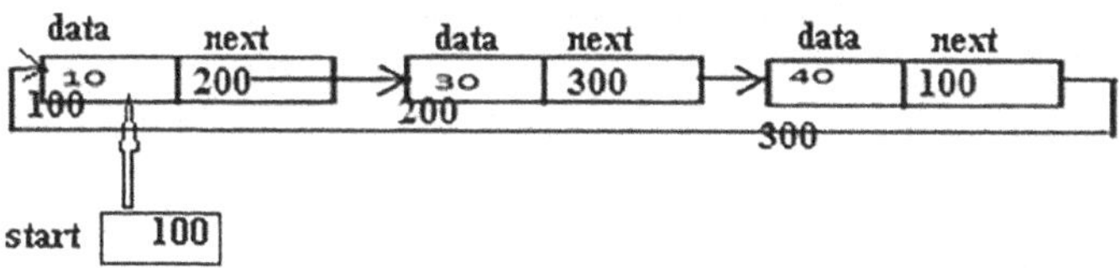

FIGURE 4.12
Singly circular linked list containing three nodes.

```
do
{
printf("%d",temp->data);
temp=temp->next;
}while(temp!=start);
```

4.5.1.3 Delete Element in Circular Linked List

Step 1: If there is only one node in a circular linked list and we want to delete the same, then we simply check whether the node's data and node which we want to delete are same or not. If it is same, then delete the node by using free function and make start=NULL (Figure 4.13);

Step 2: For node that we want to delete is not there in the linked list, display the message "No such element present in list".

Step 3: If a circular linked list contains more than one node in a circular linked list. Then set a variable found=0, if we find that node, set found=1.

Step 4: If the deleted node is the first node in your circular linked list, then apply the following steps (Figure 4.14):

Step 5: If the deleted node is not the first node in your circular linked list, then apply the following steps (Figure 4.15):

4.5.2 Program: Implementation of Singly Circular Linked List

```
//header files included
#include<stdio.h>
#include<stdlib.h>

//function declaration
void create_node();
```

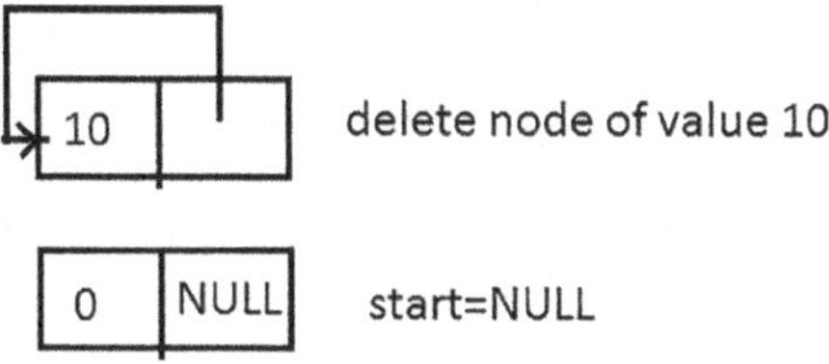

FIGURE 4.13
Delete node from circular linked list containing only one node.

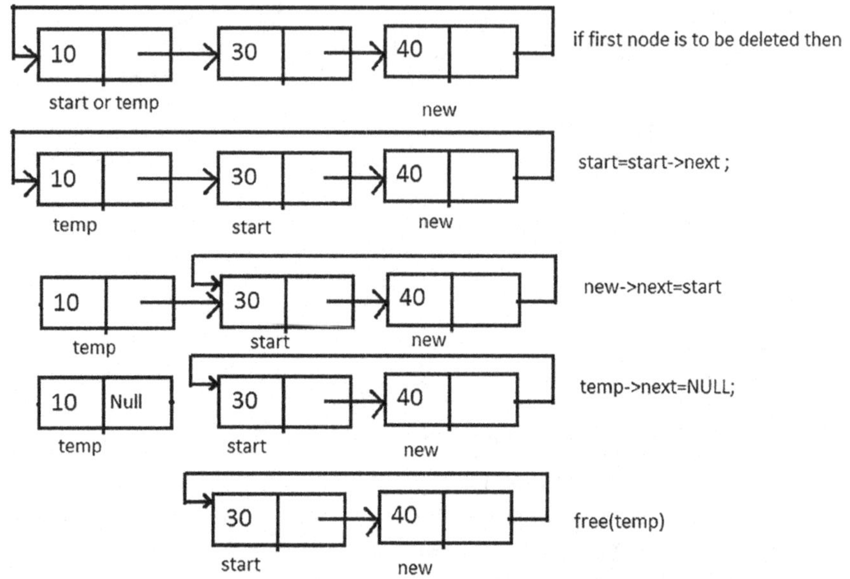

FIGURE 4.14
Delete first node from circular linked list containing three nodes.

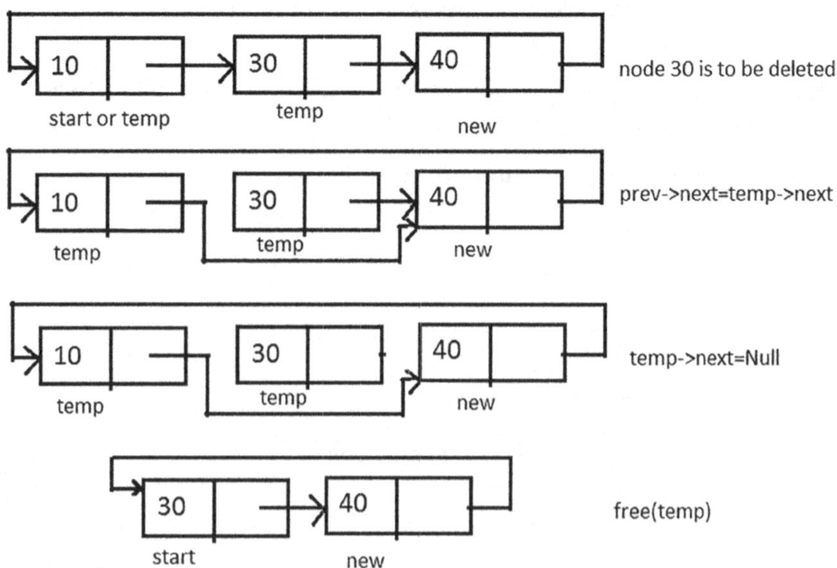

FIGURE 4.15
Delete any node other than first node from circular linked list containing three nodes.

```
void delete_node();
void display();

// Structure for node in Singly Circular Linked List
struct node
{
 int data;
 struct node *next;
}*new1,*start,*temp,*prev;
int flag=1;

//main function definition
int main()
{
 int choice;
 char ch;
 start=NULL;
 printf("Singly Circular Linked List: \n");
 do
 {
  printf("1.create \t 2.display \t 3.delete \t 4.exit: ");
  printf("\n Enter choice: ");
  scanf("%d", &choice);
      switch(choice)
      {
      case 1:create_node();
      break;
      case 2: display();
      break;
      case 3:delete_node();
      break;
      case 4:exit(0);
      default: printf("\n wrong choice");
      }
  fflush(stdin);
  printf("Do you want to continue[y/n]: ");
  scanf("%c",&ch);
  }while(ch=='y');
  return 0;
  }

//create a node in Singly Circular Linked List
void create_node()
{
 //allocate memory for node dynamically on heap memory
partition
 new1=(struct node *)malloc(sizeof(struct node));
 printf("Enter data: ");
 scanf("%d", &new1->data);
 new1->next=NULL;
 //If Linked List is empty
if(start==NULL)           // or  if(flag==1)
```

```
{
start=new1;
new1->next=start;
flag=0;
}
else
{
temp=start;
while(temp->next!=start)
{
temp=temp->next;
}
temp->next=new1;
new1->next=start;
}
printf("Node inserted at the end of Linked list=%d
\n",new1->data);
}

//display elements in Singly Circular Linked List
void display()
{
temp=start;
if(start==NULL)
{
printf("\n Singly Circular Linked List is empty.\n");
}
else
{
printf("Data in the Singly Circular Linked List is as follows:
\n");
do
{
printf("\t%d",temp->data);
temp=temp->next;
}while(temp!=start);
printf("\n");
}
}

//delete node from Singly Circular Linked List
void delete_node()
{
int num, found=0;
if(start==NULL)     //Linked list is not created
 {
  printf("\n Linked list is not created ,first create a node.
\n");
 }
 else
 {
```

```
temp=start;
printf("Enter data which you delete: ");
scanf("%d",&num);
//Singly Circular Linked List contain only one node and
deleted node's value is matched with //entered value
if((start->next==start)&&(num==start->data))
{
temp->next=NULL;
start=NULL;
free(temp);
flag=1;
return;
}
//Singly Circular Linked List contain only one node and
deleted node's value is not matched //with entered value
if((start->next==start)&&(num!=start->data))
{
printf("\n Node is not in the Singly Circular Linked
List.\n");
return;
}
 //when list contain more than one node then search element
which we want to delete and //store that node's address into
temp pointer.
do
{
if(temp->data==num)
{
found=1;
}
else
{
temp=temp->next;
}
}while((found==0)&&(temp!=start));
if(found==0)      // If node not found
{
printf("\n Node is not in the Singly Circular Linked List.\n
");
}
else        //If node is found
{
//If found node is first node in Singly Circular Linked List
then
if(temp==start)
{
start=start->next;
new1->next=start;          //Here new1 contains address of last
node in the Singly Circular Linked temp->next=NULL;        //
List because new node inserted at the end of linked list
free(temp);
```

```
temp=NULL;
}
// If found node is other than first node in Singly Circular
Linked List then
else
{
prev=start;       //prev pointer stores address of first node
initially
while(prev->next!=temp)     // after execution of while loop
prev pointer stores address of
{                                                        //previous
node which we want to delete
prev=prev->next;
}
if(temp == new1)    // if node deleted is last node in linked
list then store the address of prev
{           //pointer into new1 because new1 pointer contains
address of last node
new1= prev;
}
prev->next=temp->next;
temp->next=NULL;
free(temp);
temp=NULL;
}
printf("Node deleted =%d \n",num);
}
}
}
```

Output:

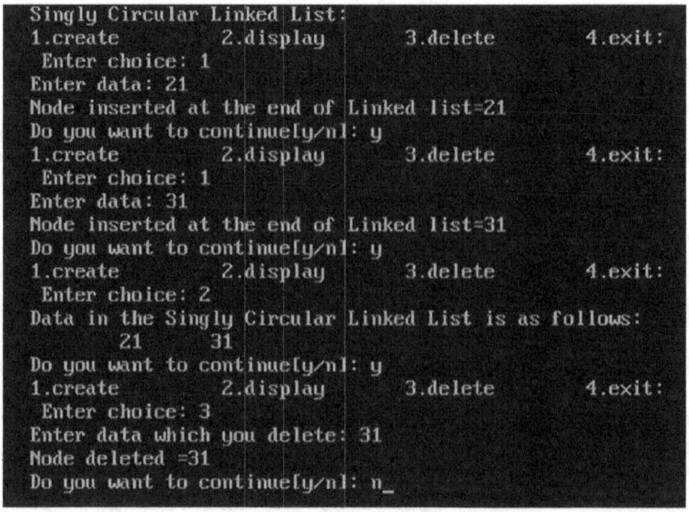

```
Singly Circular Linked List:
1.create          2.display        3.delete         4.exit:
 Enter choice: 1
Enter data: 21
Node inserted at the end of Linked list=21
Do you want to continue[y/n]: y
1.create          2.display        3.delete         4.exit:
 Enter choice: 1
Enter data: 31
Node inserted at the end of Linked list=31
Do you want to continue[y/n]: y
1.create          2.display        3.delete         4.exit:
 Enter choice: 2
Data in the Singly Circular Linked List is as follows:
        21      31
Do you want to continue[y/n]: y
1.create          2.display        3.delete         4.exit:
 Enter choice: 3
Enter data which you delete: 31
Node deleted =31
Do you want to continue[y/n]: n_
```

4.6 Doubly Linear Linked List (DLLL)

The typical structure of each node in doubly linked list is (Figure 4.16):

Doubly linked list representation using C language:

```
struct node
{
struct node *prev;
int data;
struct node *next;
};
```

Linked representation of a doubly linear linked list is as follows:

Thus, the doubly linear linked list can traverse in both the directions forward as well as backward.

4.6.1 Algorithm for Doubly Linear Linked List

4.6.1.1 Creation of Doubly Linear Linked List

Step 1: Create new node Figures 4.17 and 4.18

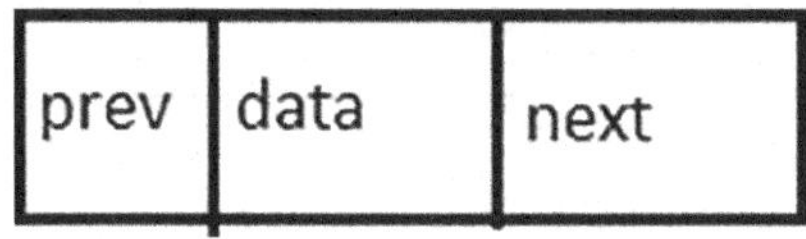

FIGURE 4.16
Node in doubly linear linked list.

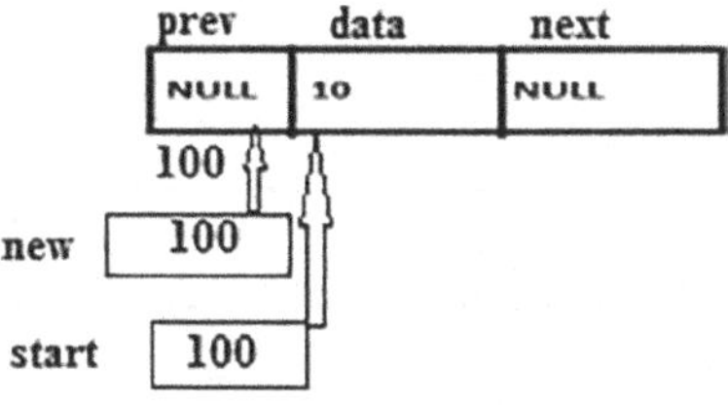

FIGURE 4.17
Doubly linear Linked list contains only one node.

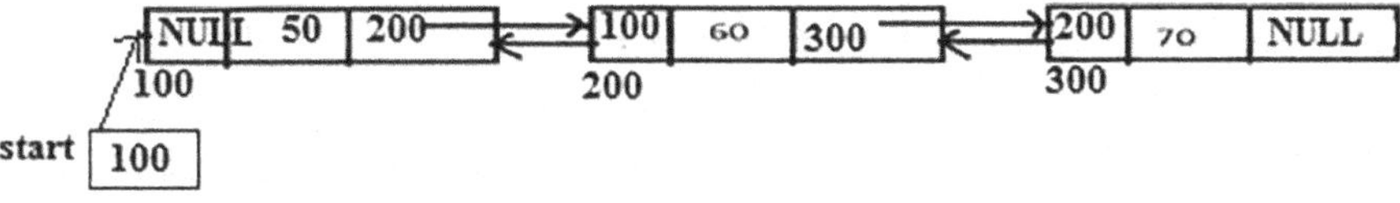

FIGURE 4.18
Doubly linear linked list with three nodes.

- Initially flag=0 or start=NULL;
- As soon as first node gets created, we reset flag=1 or first node contains address of the first node in doubly linear linked list.

Step 2: For further addition of nodes, a new node is created and inserted at the end of the list (Figure 4.19).

Step 3: For further addition, a new node is created and inserted at the end of the list as shown below.

First, search the last node in the list using code and store the address of the last node in temp pointer (Figure 4.20):

```
while( temp->next != NULL)
{
temp=temp->next;
}
```

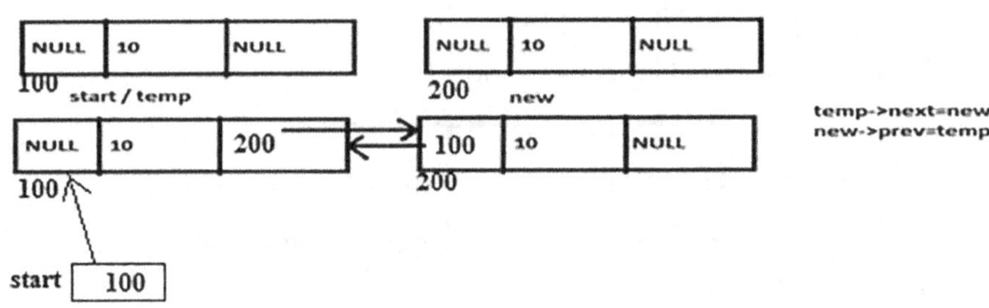

FIGURE 4.19
Insert second node in doubly linear linked list at the end.

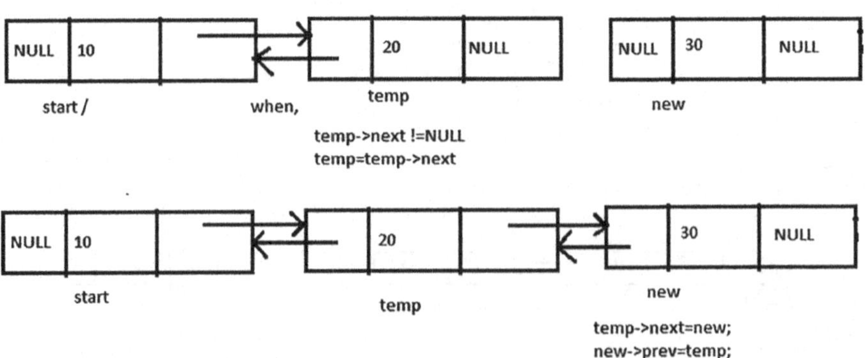

FIGURE 4.20
Insert third node in doubly linear linked list at the end.

4.6.1.2 Delete an Element from Doubly Linear Linked List

Step 1: We assume doubly linear linked list as shown below:

Step 2: If we want to delete first node in the doubly linear linked list, then here in below figure node with data 10 is to be deleted by using the following code:

Temp pointer contains the address of the node which we want to delete and temp and start addresses are same then node which we want to delete is the first node in the doubly linear linked list (Figure 4.21).

```
if(temp==start)
   {
    start=start->next;
    temp->next=NULL;
    start->prev=NULL;
    free(temp);
   }
```

Step 3: If we want to delete node other than the first node, then, here suppose node 20 is to be deleted using the following code:

temp pointer contains the address of the node which we want to delete.

If temp pointer contains the address other than first pointer, then node which we want to delete is other than first node in the doubly linear linked list.

Then, assign the address of temp node's next address to the temp node's previous node's next field using, **(temp->prev)->next=temp->next;**

Then, assign the address of temp node's previous address to the temp node's next node's previous field using, **(temp->next)->prev=temp->prev;**

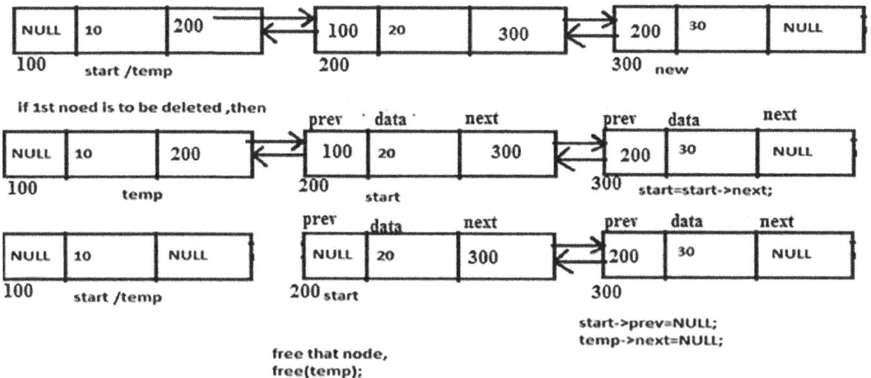

FIGURE 4.21
Delete a start node from doubly linear linked list.

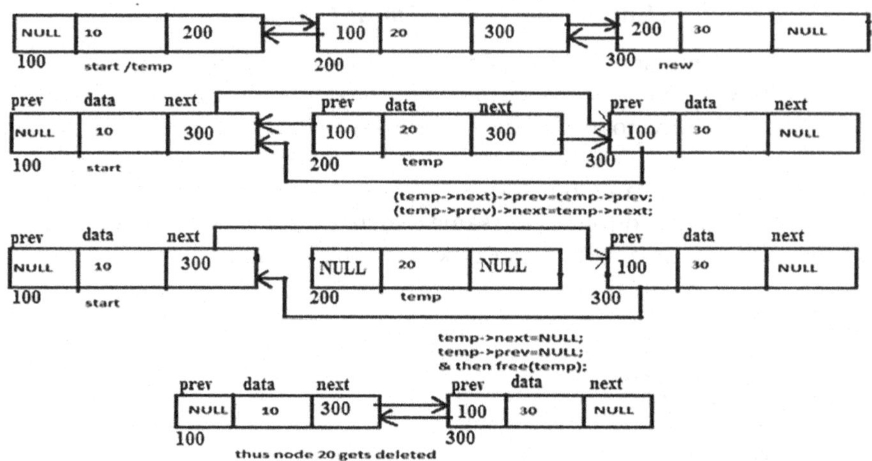

FIGURE 4.22
Delete any node other than a start node from doubly linear linked list.

Then remove the links between temp and next node and temp and previous node using,

```
temp->next=NULL;
temp->prev=NULL;
```

Then de-allocate memory that is reserved for temp node using,

```
free(temp);
```

Here, dynamically reserved memory on heap get de-allocated and assign temp=NULL means temp pointer does not point to de-allocated location and dangling pointer problem is resolved (Figure 4.22).

```
temp=NULL ;
```

4.6.2 Program: Implementation of Doubly Linear Linked List

```
#include<stdio.h>
#include<stdlib.h>
struct node
{
 struct node *prev;
 int data;
 struct node *next;
}*new1,*start,*temp;
int flag=1;
//create a node and insert at end of doubly linear linked
list.
```

```
void create_node()
{
 //allocate dynamic memory for a node on heap partition on RAM
 new1=(struct node *)malloc(sizeof(struct node));
 printf("Enter data: ");
 scanf("%d",&new1->data);
 new1->next=NULL;
 new1->prev=NULL;
 // doubly linear linked list is empty
 if(start==NULL)           // or  if(flag==1)
 {
  start=new1;
  flag=0;
 }
 // doubly linear linked list contains more than one node
else
{
temp=start;
while(temp->next!=NULL)
{
temp=temp->next;
}
temp->next=new1;
new1->prev=temp;
}
printf("Node inserted at the end of Linked list=
%d\n",new1->data);
}
//display data part in doubly linear linked list
void disp()
{
temp=start;
// doubly linear linked list is empty
if(start==NULL)
{
printf("\n Doubly linear linked list is empty.\n");
return;
}
else
{
printf("Data in the Doubly Linear Linked List is as follows:
\n");
while(temp!=NULL)
{
printf("\t%d",temp->data);
temp=temp->next;
}
printf("\n");
}
}
```

```
//delete a node from the doubly linear linked list.
void delete_node()
{
 int num;
 int found=0;
 temp=start;
 // doubly linear linked list is empty
 if(start==NULL)
 {
  printf("\n Doubly linear linked list is empty. \n");
  return;
 }
 // doubly linear linked list is created and having some nodes
 else
 {
  printf("Enter data which you want to delete: ");
  scanf("%d",&num);
  //search element which we want to delete and store address
of node which we want to delete
  //into temp pointer
  while((found==0)&&(temp!=NULL))
  {
   if(temp->data==num)
   {
       found=1;
   }
   else
   {
       temp=temp->next;
   }
  }
  //Data element is not present in the doubly linear linked
list
  if(found==0)
  {
   printf("\n Node not found. \n");
   return;
  }
  //Data element is present in the doubly linear linked list
  else
  {
   //Node which we want to delete is first node in the doubly
linear linked list
    if(temp==start)
    {
       if(temp->next!=NULL) //doubly linear linked list
contains more than one node and we want to delete is first
node
{
      start=start->next;
      temp->next=NULL;
```

```
      start->prev=NULL;
      free(temp);
      temp=NULL;
      }else //doubly linear linked list contains only one node
      and we want to delete is first node
      {
      free(temp);
      start=NULL;
      }
   }
   //Node which we want to delete is other than first node in
the doubly linear linked list
 else
   {
       if(temp->next!=NULL) // Node deleted is not last node in
the doubly linear linked list
       {
            (temp->prev)->next=temp->next;
            (temp->next)->prev=temp->prev;
       }else   // Node deleted is last node in the doubly
linear linked list
       {
       (temp->prev)->next=NULL;
       }

             temp->next=NULL;
             temp->prev=NULL;
             free(temp);
             temp=NULL;
   }
 printf("Node deleted =%d \n",num);
 }
 }
 }   // end of delete_node function
//main function
int main()
{
 int choice;
 char ch;
 start=NULL;
 printf("Doubly linear linked list: ");
 do
 {
  printf("1. create\t2. display\t3. delete\t4. exit");
  printf("\n Enter your choice: ");
  scanf("%d",&choice);
  switch(choice)
  {
   case 1:create_node();
   break;
   case 2:disp();
```

```
   break;
   case 3:delete_node();
   break;
   case 4:exit(0);
   default:printf("\n Wrong choice");
  }
  fflush(stdin);
  printf("Do you want to continue[y/n]");
  scanf("%c",&ch);
 }while(ch=='y');
 return 0;
}
```

Output:

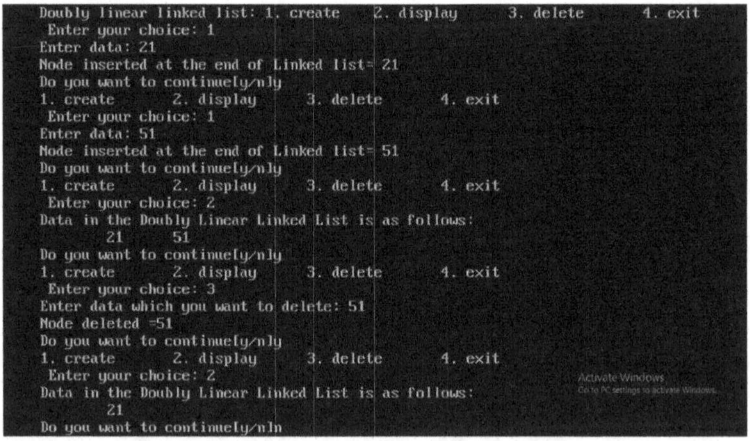

```
Doubly linear linked list: 1. create    2. display    3. delete    4. exit
 Enter your choice: 1
Enter data: 21
Node inserted at the end of Linked list= 21
Do you want to continue[y/n]y
1. create       2. display      3. delete       4. exit
 Enter your choice: 1
Enter data: 51
Node inserted at the end of Linked list= 51
Do you want to continue[y/n]y
1. create       2. display      3. delete       4. exit
 Enter your choice: 2
Data in the Doubly Linear Linked List is as follows:
        21      51
Do you want to continue[y/n]y
1. create       2. display      3. delete       4. exit
 Enter your choice: 3
Enter data which you want to delete: 51
Node deleted =51
Do you want to continue[y/n]y
1. create       2. display      3. delete       4. exit
 Enter your choice: 2
Data in the Doubly Linear Linked List is as follows:
        21
Do you want to continue[y/n]n
```

4.6.3 Insert New Node at the Start of Doubly Linear Linked List

```
void insert_at_start()
{
      new1=(struct node*)malloc(sizeof(struct node));
      printf("\n Enter data in new node:");
      scanf("%d",&new1->data);
      new1->next=NULL;
      new1->prev=NULL;
      if(start==NULL)
{
      start=new1;
}
      else
{
                        new1->next=start;
                        start->prev=new1;
```

```
                        start=new1;
      }
display(); // display all the node data from start node to
end of linked list.
}
```

4.6.4 Insert New Node at Last Position of Doubly Linked List

```
void insert_at_last()
{
      new1=(struct node*)malloc(sizeof(struct node));
      printf("\nEnter data in node:");
      scanf("%d",&new1->data);
      new1->next=NULL;
      new1->prev=NULL;
      if(start==NULL)
      {
                start=new1;
      }
      else
      {
                temp=start;
                while(temp->next!=NULL)
                {
                temp=temp->next;
                }
                temp->next=new1;
                new1->prev=temp;
      }
display(); // display all the node data from start node to end
of linked list.
}
```

4.6.5 Insert New Node at the Specified Position in Doubly Linear Linked List

```
void insert_at_specificpos()
{
int pos, count=0, i=1;
new1=(struct node*)malloc(sizeof(struct node));
printf("\n Enter data in node:");
scanf("%d", &new1->data);
new1->next=NULL;
new1->prev=NULL;
if(start==NULL)
{
start=new1;
}
else
```

```
{
temp=start;
while(temp!=NULL)    //count number of nodes present in your
linked list
{
temp=temp->next;
count++;
}
lable1:
printf("\n Enter position to insert new node:"); //position
must be greater than zero
scanf("%d",&pos);
if( pos > (cont+1))
{
printf("\n Your choice is wrong");
goto lable1;
}
if(pos==1)     //if node inserted at first position in doubly
linear linked list
{
start->prev=new1;
new1->next=start;
start=new1;
start->prev=NULL;
}
else                    // if node inserted at other than first
position in doubly linear linked list
{
temp=start;
while(i < (pos-1))  //traverse the temp pointer till temp
pointer contains address of previous
{                              //node where we want to insert
the new node
temp=temp->next;
i++;
}
new1->next=temp->next;
temp->next=new1;
new1->prev=temp;
}
printf("\n Doubly linear linked list data:");
temp=start;
while(temp!=NULL)  //display all nodes data from first node to
last node
{
printf("%d\t", temp->data);
temp=temp->next;
}
}
}
```

4.7 Doubly Circular Linked List

Doubly circular linked list contains two pointers previous and next pointers. Previous pointer points to the previous node and next pointer points to the next node or contains the address of the next node. Last node's next pointer stores the address of the first node and the first node's previous pointer stores the address of the last node in the doubly circular linked list.

4.7.1 Program: Implementation of Doubly Circular Linked List

```
#include<stdio.h>
#include<stdlib.h>
struct node
{
int data;
struct node* prev;
struct node* next;
}*start,*new1,*temp,*prev,*last;
//last pointer stores address of last node in Doubly Circular
Linked List
void insert_at_end();
void display();
void reverse();
void delet();
//main function
int main()
{
int c;
char ch;
printf("Doubly circular linked list: \n");
do
{
printf("1.Insert_at_end  2.Display nodes in reverse  3. delete
node  4. display");
printf("\n Enter your choice: ");
scanf("%d", &c);
switch(c)
{
case 1:
      insert_at_end();
      break;
case 2:
      reverse();
      break;
case 3:
      delet();
      break;
```

```
case 4:
      display();
      break;
default:
      printf("\n You have entered wrong choice.\n");
}
fflush(stdin);
printf("Do you want to continue(Y/N): ");
scanf("%c",&ch);
}while(ch=='Y'||ch=='y');
return 0;
}
void display()
{
if(start==NULL)
{
printf("\n Doubly Circular Linked List is not created.\n");
return;
}
else
{
printf("Doubly Circular Linked List data:");
temp=start;
do
{
printf("%d\t",temp->data);
temp=temp->next;
} while(temp!=start);
printf("\n");
}
}
void insert_at_end()
{
new1=(struct node*)malloc(sizeof(struct node));
printf("Enter data in node: ");
scanf("%d",&new1->data);
new1->next=NULL;
new1->prev=NULL;
if(start==NULL)
{
start=new1;
start->next=start;
start->prev=start;
}
else
{
temp=start;
do
{
temp=temp->next;
```

```
} while(temp->next!=start);
temp->next=new1;
new1->prev=temp;
new1->next=start;
start->prev=new1;
}
display();
}
//display data in the doubly circular linked list in reverse
order
void reverse()
{
if(start==NULL)
{
printf("\n Doubly Circular Linked List is not created.");
return;
}
else
{
temp=start;
do                // Using do while loop we have to move temp
pointer at the last node
{        //Here use do while or while loop both works
correctly
temp=temp->next;
} while(temp->next!=start);
// Now after exit from do while loop temp pointer contains
address of last node of doubly //circular linked list
do
{
printf("\t%d",temp->data);
temp=temp->prev;
}while(temp!=start->prev);     //display data field from last
node to first node in reverse order
printf("\n");
}
}
void delet()
{
int num,found=0;
if(start==NULL)
{
printf("Doubly Circular Linked List is not created.\n");
return;
}
else
{
printf("Enter value of node which you want to delete: ");
scanf("%d",&num);
```

```
//If linked list contains only one node and that matched with
node which we want to delete
if(start==start->next && num==start->data)
{
start->next=NULL;
start->prev=NULL;
free(start);
start=NULL;
printf("Node deleted is %d\n",num);
return;
}
//If linked list contains only one node and that not matched
with node which we want to //delete
if(start==start->next && num!=start->data)
{
printf("\n Data not present in Doubly Circular Linked List ");
return;
}
temp=start;
do
{
if(temp->data==num)
{
found=1;
}
else
{
temp=temp->next;
}
}while(found!=1 && temp!=start);
if(found==0)   //if node is not found in Doubly Circular
Linked List
{
printf("\n Node is not present in Doubly Circular Linked List
");
return;
}
else       //if node is found in Doubly Circular Linked List
{
if(temp==start)   //if first node we want to delete from
Doubly Circular Linked List
{
last=start;
do
{
last=last->next;
}
while(last->next!=start);
start=start->next;
last->next=start;
```

```
start->prev=last;
temp->next=NULL;
temp->prev=NULL;
free(temp);
temp=NULL;
}
else   //if other than first node we want to delete from
Doubly Circular Linked List
{
(temp->prev)->next=temp->next;
(temp->next)->prev=temp->prev;
temp->next=NULL;
temp->prev=NULL;
free(temp);
temp=NULL;
}
printf("Node deleted is %d\n",num);
}
}
}
```

Output:

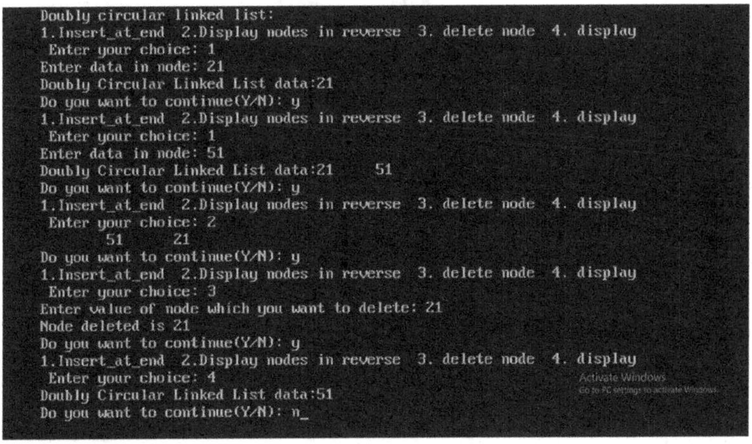

```
Doubly circular linked list:
1.Insert_at_end  2.Display nodes in reverse  3. delete node  4. display
 Enter your choice: 1
Enter data in node: 21
Doubly Circular Linked List data:21
Do you want to continue(Y/N): y
1.Insert_at_end  2.Display nodes in reverse  3. delete node  4. display
 Enter your choice: 1
Enter data in node: 51
Doubly Circular Linked List data:21     51
Do you want to continue(Y/N): y
1.Insert_at_end  2.Display nodes in reverse  3. delete node  4. display
 Enter your choice: 2
        51      21
Do you want to continue(Y/N): y
1.Insert_at_end  2.Display nodes in reverse  3. delete node  4. display
 Enter your choice: 3
Enter value of node which you want to delete: 21
Node deleted is 21
Do you want to continue(Y/N): y
1.Insert_at_end  2.Display nodes in reverse  3. delete node  4. display
 Enter your choice: 4
Doubly Circular Linked List data:51
Do you want to continue(Y/N): n_
```


4.8 Stack Implementation Using Linked List

4.8.1 Representation of Stack Using Linked List

- The advantage of implementing stack using a linked list is that we need not have to worry about the size of the stack.
- Nodes are dynamically created so there will not be any stack full condition.

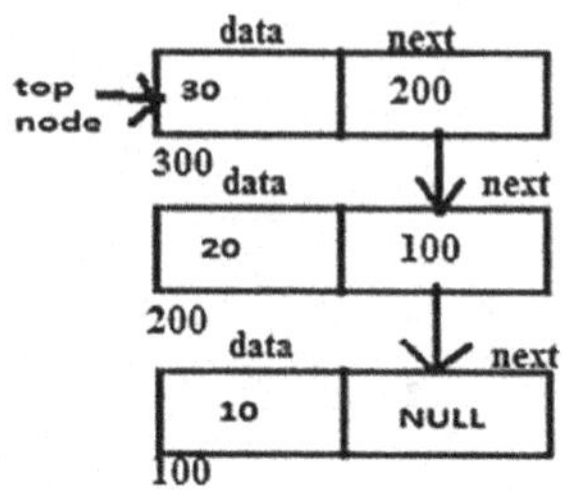

FIGURE 4.23
Stack representation using linked list.

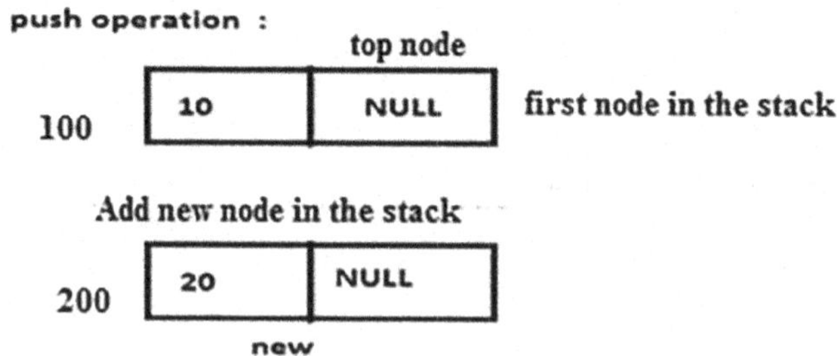

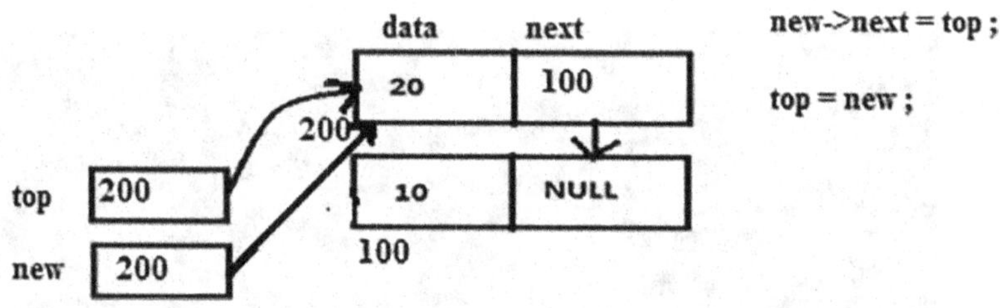

FIGURE 4.24
Push operation on stack using linked list.

C structure for stack using linked list (Figures 4.23 and 4.24):

```
struct stack
{
      int data;
      struct stack *next;
}*new1,*top,*temp;
```

4.8.2 Program: Stack Implementation Using Linked List

```
#include<stdio.h>
#include<stdlib.h>
struct stack
```

```
{
int data;
struct stack *next;
}*new1,*top,*temp;
// top pointer stores address of top node in your stack.
// temp pointer stores temporary address of a node and used
for traversing purpose.
// new1 pointer stores the address of newly created node of
your stack.
void push();
int pop();
void display();
int main()
{
int ch, item;
char c;
top =NULL;
do
{
printf(" Stack implementation using linked list: \n");
printf(" 1.push 2.pop 3.display 4.exit");
printf("\n Enter your choice: ");
scanf("%d",&ch);
switch(ch)
{
case 1:
push();
break;
case 2:
item=pop();
printf("Popped element from the stack=%d \n", item);
break;
case 3:
display();
break;
case 4:
exit(0);
default:
printf("\n wrong choice\n");
}
fflush(stdin);
printf("Do you want to continue(y/n):");
scanf("%c",&c);
} while(c=='y'||c=='Y');
return 0;
}
void push()
{
new1=(struct stack*)malloc(sizeof(struct stack));
printf("Enter data: ");
```

```
scanf("%d",&new1->data);
new1->next=NULL;
if(top == NULL)
{
top=new1;
}else
{
new1->next = top;
top = new1;
}
printf("Pushed element on stack is %d \n",new1->data);
}
int pop()
{
int data;
temp= top;
if(top==NULL)
{
printf("stack is underflow\n");
}
else
{
data= top ->data;
top = top ->next;
temp->next=NULL;
free(temp);
temp = NULL;
return(data);
}
}
void display()
{
temp=top;
if(top ==NULL)
{
printf("\n stack is empty");
}
else
{
while(temp!=NULL)
{
printf("\t %d",temp->data);
temp=temp->next;
}
printf("\n");
}
}
```

Output:

```
 Stack implementation using linked list:
 1.push 2.pop 3.display 4.exit
 Enter your choice: 1
Enter data: 21
Pushed element on stack is 21
Do you want to continue(y/n):y
 Stack implementation using linked list:
 1.push 2.pop 3.display 4.exit
 Enter your choice: 1
Enter data: 31
Pushed element on stack is 31
Do you want to continue(y/n):y
 Stack implementation using linked list:
 1.push 2.pop 3.display 4.exit
 Enter your choice: 3
        31      21
Do you want to continue(y/n):y
 Stack implementation using linked list:
 1.push 2.pop 3.display 4.exit
 Enter your choice: 2
Popped element from the stack=31
Do you want to continue(y/n):y
 Stack implementation using linked list:
 1.push 2.pop 3.display 4.exit
 Enter your choice: 4
```

4.9 Linear Queue Implementation Using Linked List

4.9.1 Representation of Queue Using Linked List

The main advantage in linked representation is that we need not have to worry about the size of the queue. Here, there will not be a queue full condition at all. The queue using the linked list will be very much similar to a linked list. The only difference between the two is in queue left most node is called front node and rightmost node is called rear node. We cannot remove any arbitrary node from it, and we always have to remove front node.

C structure for linear queue using linked list (Figure 4.25):

```
struct queue
{
int data;
struct queue *next;
}*new1,*rear,*front,*temp;
```

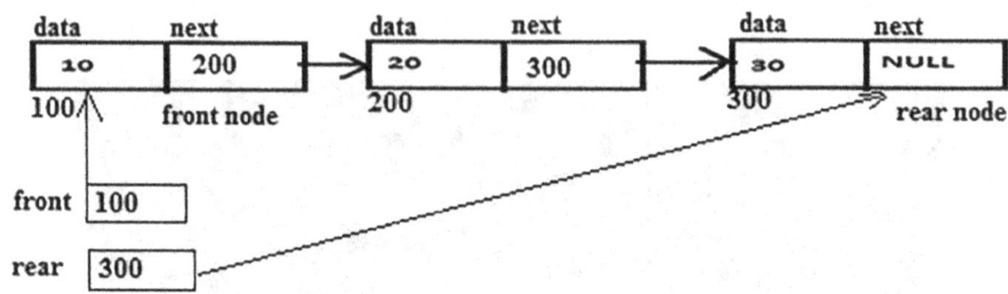

FIGURE 4.25
Linear queue representation using linked list.

```
//front pointer store address of front or first node in linear
queue
//rear pointer store address of rear or last node in linear
queue
//new1 pointer store address of newly created node
//temp pointer store address of temporary node and used for
traversing purpose
```

4.9.2 Program: Linear Queue Implementation Using Linked List

```
#include<stdio.h>
#include<stdlib.h>
//structure of linear queue using linked list
struct queue
{
 int data;
 struct queue *next;
}*new1,*rear,*front,*temp;
//function prototypes or function declaration
void create();
void delet();
void display();

//main() function
int main()
{
 char ch;
 int choice;
 front=rear=NULL;
 do
 {
  printf("Linear queue using linked list: ");
  printf("\n 1.create 2.delete 3.display 4.exit: ");
  printf("\n Enter choice: ");
```

```
  scanf("%d", &choice);
  switch(choice)
  {
   case 1:
create();
      break;
   case 2:
delet();
      break;
   case 3:
display ();
      break;
   case 4:
exit(0);
   default:
printf("\n wrong choice");
  }
  fflush(stdin);
  printf("Do you want to continue...[y/n]");
  scanf("\n%c",&ch);
 }while(ch=='y' || ch=='Y');
 return 0;
}

//create linear queue using linked list that is enQueue
operation
void create()
{
new1=(struct queue *)malloc(sizeof(struct queue));
printf("Enter a data: ");
scanf("%d",&new1->data);
new1->next=NULL;
if(front==NULL)
{
front=new1;
rear=new1;
}
else
{
rear->next=new1;
rear=new1;
}
printf("\n Inserted element in linear queue =
%d\n",new1->data);
}
//delete an element from a linear queue that is deQueue
operation
void delet()
{
```

```
temp=front;
if(front==NULL)
{
printf("\n Linear queue is empty");
return;
}
 else
 {
if( front == rear)
{
printf("Deleted element = %d \n", front->data);
front = NULL;
rear = NULL;
free(temp);
temp=NULL;
}else
{
printf("Deleted element = %d \n", front->data);
      front=front->next;
      temp->next=NULL;
      free(temp);
      temp=NULL;
}
}
}
//display linear queue elements from the front end to rear end
void display ()
{
temp=front;
if(front==NULL)
{
printf("\n Queue is empty");
}
else
{
printf("Data in linear queue is as follows: \n");
while(temp!=NULL)
{
printf("\t%d",temp->data);
temp=temp->next;
}
printf("\n");
}
}
```

Output:

```
Linear queue using linked list:
 1.create 2.delete 3.display 4.exit:
 Enter choice: 1
Enter a data: 21

 Inserted element in linear queue = 21
Do you want to continue...[y/n]y
Linear queue using linked list:
 1.create 2.delete 3.display 4.exit:
 Enter choice: 1
Enter a data: 51

 Inserted element in linear queue = 51
Do you want to continue...[y/n]y
Linear queue using linked list:
 1.create 2.delete 3.display 4.exit:
 Enter choice: 3
Data in linear queue is as follows:
        21      51
Do you want to continue...[y/n]y
Linear queue using linked list:
 1.create 2.delete 3.display 4.exit:
 Enter choice: 2
Deleted element = 21
Do you want to continue...[y/n]y
Linear queue using linked list:
 1.create 2.delete 3.display 4.exit:
 Enter choice: 3
Data in linear queue is as follows:
        51
Do you want to continue...[y/n]n
```

4.10 Circular Queue Implementation Using Linked List

4.10.1 Operations on Circular Queue

Front pointer stores the address of the front node in circular queue.

Rear pointer stores the address of the rear or last node in circular queue.

1. **enQueue operation:** The purpose of this operation or function is to insert an item into the circular queue. In a circular queue, the new item is always placed in the rear position.

Steps:

1. Create a new node dynamically and insert the data into the data fields and next contains the address of NULL initially.
2. Verify if front==NULL, if it is true, then front=rear=address of the newly created node.
3. If it is false, then rear=address of the newly created node and front node always contains the address of the front node.

2. **Dequeue operation:** This operation or function is used to remove an element from the circular queue. In a queue, the element is always deleted from the front position.

   Steps:

   1. Check whether the queue is empty or not means front == NULL.
   2. If it is empty then displaying queue is empty. If the queue is not empty, then GOTO step 3.
   3. Check if (front==rear) if it is true, then circular queue contains only one node, then set front=rear=NULL else move the front forward in the queue that is front=front->next. Display the deleted element.

C structure for circular queue using linked list (Figure 4.26):

```
struct cqueue
{
int data;
struct cqueue *next;
}*new1,*rear,*front,*temp;
//front pointer stores the address of front or first node in
circular queue
//rear pointer stores the address of rear or last node in
circular queue
//new1 pointer stores the address of newly created node
//temp pointer stores the address of temporary node and used
for traversing purpose
```

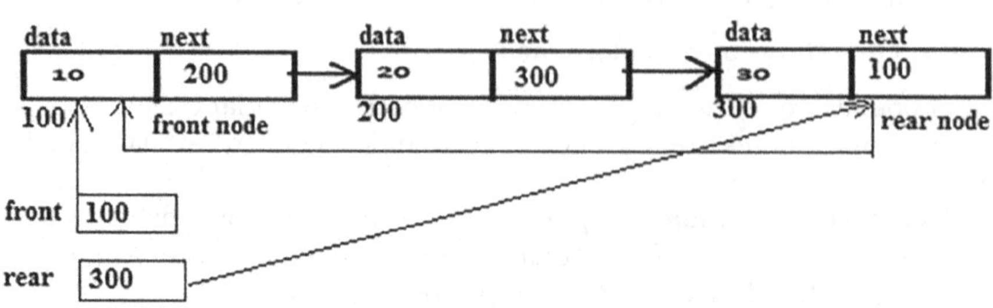

FIGURE 4.26
Circular queue representation using linked list.

4.10.2 Program: Circular Queue Implementation Using Linked List

```
#include<stdio.h>
#include<stdlib.h>
//structure of circular queue using linked list
struct cqueue
{
 int data;
 struct cqueue *next;
}*new1,*rear,*front,*temp;
//function prototypes or function declaration
void create();
void delet();
void display();
//main() function
int main()
{
 char ch;
 int choice;
 front=rear=NULL;
 do
 {
  printf("Circular Queue using  linked list: ");
  printf("1.create 2.delete 3.display 4.exit");
  printf("\n Enter choice: ");
  scanf("%d", &choice);
         switch(choice)
{
         case 1:
  create();
      break;
      case 2:
  delet();
      break;
      case 3:
  display ();
      break;
      case 4:
  exit(0);
      default:
  printf("\n wrong choice:");
  }
fflush(stdin);
printf("Do you want to continue...[y/n]");
scanf("%c", &ch);
}while(ch=='y' || ch=='Y');
return 0;
}
```

```
//create circular queue using linked list that is enQueue
operation
void create()
{
new1=(struct cqueue *)malloc(sizeof(struct cqueue));
printf("Enter a data: ");
scanf("%d",&new1->data);
new1->next=NULL;
if(front==NULL)
{
front=new1;
rear=new1;
rear->next = front;
}
else
{
rear->next=new1;
rear=new1;
rear->next = front;
}
printf("Inserted element in circular queue =
%d\n",new1->data);
}

//delete an element from a circular queue that is deQueue
operation
void delet()
{
temp=front;
if(front==NULL)
{
printf("\n Circular Queue is empty");
return;
}
 else
 {
 if(front == rear)
{
printf("Deleted element = %d\n", front->data);
front = NULL;
rear = NULL;
free(temp);
temp=NULL;
}else
{
printf("Deleted element = %d\n", front->data);
temp = front;
front = front->next;
temp->next=NULL;
free(temp);
rear->next = front;
```

```
temp=NULL;
}
}
}

//display circular queue elements from the front end to rear
end
void display ()
{
temp=front;
if(front==NULL)
{
printf("Circular queue is empty\n");
}
else
{
printf("Data in circular queue is as follows: \n");
do
{
printf("\t%d",temp->data);
temp=temp->next;
}while(temp != front);
printf("\n");
}
}
```

Output:

```
Circular Queue using  linked list: 1.create 2.delete 3.display 4.exit
 Enter choice: 1
Enter a data: 21
Inserted element in circular queue = 21
Do you want to continue...[y/n]y
Circular Queue using  linked list: 1.create 2.delete 3.display 4.exit
 Enter choice: 1
Enter a data: 71
Inserted element in circular queue = 71
Do you want to continue...[y/n]y
Circular Queue using  linked list: 1.create 2.delete 3.display 4.exit
 Enter choice: 2
Deleted element = 21
Do you want to continue...[y/n]y
Circular Queue using  linked list: 1.create 2.delete 3.display 4.exit
 Enter choice: 3
Data in circular queue is as follows:
        71
Do you want to continue...[y/n]y
Circular Queue using  linked list: 1.create 2.delete 3.display 4.exit
 Enter choice: 1
Enter a data: 99
Inserted element in circular queue = 99
Do you want to continue...[y/n]y
Circular Queue using  linked list: 1.create 2.delete 3.display 4.exit
 Enter choice: 3
Data in circular queue is as follows:
        71      99
Do you want to continue...[y/n]y
Circular Queue using  linked list: 1.create 2.delete 3.display 4.exit
 Enter choice: 4
```

4.11 Interview Questions

1. What is a linked list?
2. What are the advantages of linked list over array?
3. Can we apply binary search algorithm to a sorted linked list, why?
4. What is the difference between a stack and an array?
5. What are the disadvantages of array implementation over the linked list?
6. What are the disadvantages of linear linked list?
7. Define circular linked list.
8. What are the disadvantages of circular linked list?
9. Define and explain double linked list with a suitable example.
10. Explain whether linked list is linear or nonlinear data structure.
11. What are the parts of a linked list?
12. How will you represent a linear and circular linked list in a graphical view?
13. List all types of linked list. Explain circular linked list in detail.
14. Explain doubly linked list.
15. How do you search for a target key in a linked list?
16. What is a circular linked list? Explain its applications.
17. How many different pointers are necessary to implement a singly linear linked list?
18. How can we represent a doubly circular linked list node?
19. How can you insert a node to the beginning of a singly circular linked list?
20. How can we insert a node at the end of doubly circular linked list?
21. Write a procedure which adds a node at the specific position of doubly linear linked list?
22. How can we delete any specific node from the singly linear linked list?
23. Write a function which deletes first node from the singly circular linked list?
24. How can we reverse a singly circular linked list?
25. How to calculate the length of a singly linear linked list?
26. Explain the drawbacks of the linked list.
27. What are the major differences between the linked list and linear array?

28. Write a C code for circular queue implementation using linked list?
29. Write a C code for stack implementation using linked list?
30. Implement C code for linear queue using linked list with a suitable example.
31. Write C code for displaying all data members of singly linear linked list in reverse manner.
32. Write an algorithm and C code for deleting any node from the singly linear linked list.

4.12 Multiple Choice Questions

1. Which of the following statements is true regarding the data structure of the linked list when compared to the array?
   A. Arrays have better cache locality that can make them superior in terms of efficiency.
   B. Inserting, displaying and deleting items into a linked list are easy.
   C. Random access is not permitted within a typical linked lists implementation.
   D. The size of the array should be decided at the time of compilation, and linked lists may change their size at any time dynamically.
   E. **All of the above**

   Answer: (E)

2. The following C function takes a singly linked list as an input argument. It modifies the list by moving the last element to the front of the list and returns the modified list. Some part of the code is left blank.

```
typedef struct node {
        int value;
        struct node *next;
        }    Node;
Node *move_to_front(Node *head)
{
 Node *p, *q;
      if((head = = NULL || (head->next = = NULL))
           return head;
```

```
        q = NULL; p = head;
        while (p-> next !=NULL) {
                q=p;
                p=p->next;
                                        }
        return head;
}
```

Choose the correct alternative to replace the blank line

A. q=NULL; p->next=head; head=p;

B. q->next=NULL; head=p; p->next=head;

C. head=p; p->next=q; q->next=NULL;

D. **q->next=NULL; p->next=head; head=p**

Answer: (D)

1 Explanation:

1. Argument passed to the function move_to_front is head of linked list
2. Inside function it checks condition to decide list is empty or with single node. if condition is true then it returns head as it is without modification.
3. if condition is false means with more than one node then while loop iterates up to second last node with condition P-->next!=NULL. q points to the second last node and p to last node.
4. Now, we have to move last node to first,
   a. Make second last node next to NULL q-->next=NULL
   b. P becomes first node and head becomes second so p--> next should point to head.
   c. Make p as head. head=p.

So, D option is correct.

3. The following C function takes a list of integers as parameter and rearranges the elements in the list. The function is called with the list containing the 1, 2, 3, 4, 5, 6, 7 integers in the order indicated. What will be the contents of the list after the function has been executed?

```
struct node {
int value;
struct node * next;
};
Void rearrange (struct node * list) {
```

```
struct node * p, * q;
int temp;
if (!list || !list  -> next) return;
p = list; q = list  -> next;
while (q) {
temp = p  -> value; p -> value = q -> value;
q -> value = temp; p = q -> next
q = p? p -> next : 0;
}
}
```

A. 1,2,3,4,5,6,7

B. 2,1,4,3,6,5,7

C. 1,3,2,5,4,7,6

D. 2,3,4,5,6,7,1

Answer: (B)
Explanation:

4. Function rearrange takes list of integers as an input.
5. Inside function it checks condition to decide list is empty or with single node. If condition is true then it returns head as it is without modification.
6. Then, first element will get assigned to p and second element to q.
7. Inside while at first iteration value of p and q will get swapped means values of first element and second element.
8. Then P modified and points to third element.
9. If p is null then q holds 0 else q points to the 4th element
10. Means p points to odd number element and q points to even number elements and then we are swapping their values.
11. So output will be 2,1, 4, 3, 6, 5, 7.

4. What does the following operation do for a given linked list with an initial node as head?

```
void fun1(struct node* head)
{
  if(head == NULL)
    return;
  fun1(head->next);
  printf("%d  ", head->data);
}
```

A. Prints all nodes of linked lists

B. **Prints all nodes of linked list in reverse order**

C. Prints alternate nodes of Linked List

D. Prints alternate nodes in reverse order

Answer: (B)
Explanation:

1. In the above example, recursive function call is used.
2. printf statement is written after function call, so it will get executed after returning from last function call.
3. Last function call will be to call func1 with null, so it will simply return.
4. Second last function call will be with last node, so it will display last node value.
5. In this way, linked list will get displayed in reverse order.

For Linked List 1->2->3->4->5, fun1() prints 5->4->3->2->1.

5. The following function reverse() is supposed to reverse a singly linked list. There is one line missing at the end of the function.

```
/* Link list node */
struct node
{
    int data;
    struct node* next;
};
/* head_ref  is a double pointer which points to head (or
start) pointer
  of linked list */
static void reverse(struct node** head_ref)
{
    struct node* prev   = NULL;
    struct node* current = *head_ref;
    struct node* next;
    while (current != NULL)
    {
        next  = current->next;
        current->next = prev;
        prev = current;
        current = next;
    }
    /* predict the statement here */
}
```

What needs to be added instead of "/* predict the statement here */", so that the function correctly reverses a linked list.

A. *head_ref=prev;
B. *head_ref=current;
C. *head_ref=next;
D. *head_ref=NULL;

Answer: (A)
Explanation:

```
*head_ref = prev;
```

At the end of while loop, the *prev* pointer points to the last node of original linked list. We need to change *head_ref so that the head pointer now starts pointing to the last node.

6. What is the result of the following function to start pointing to the first node of follow linked list?

```
1->2->3->4->5->6
void fun(struct node* start)
{
  if(start == NULL)
    return;
  printf("%d  ", start->data);

  if(start->next != NULL )
    fun(start->next->next);
  printf("%d  ", start->data);
}
```

A. 1 4 6 6 4 1
B. 1 3 5 1 3 5
C. 1 2 3 5
D. 1 3 5 5 3 1
Answer: (D)
Explanation:
In above example recursive function call is used:

Function Call		no	1st printf	2nd printf
1	fun(start)			
	fun(pointer to 1)	1	1	
2	fun(start->next->next)	fun(pointer to 3)	3	3
3	fun(start->next->next)	fun(pointer to 5)	5	5
4	fun(NULL)			

1. First printf is written before function call so it will execute as per the order

   3 5

2. Second function call is after function call so it will execute in reverse order

   5 3 1

   Therefore, output will be

   1 3 5 5 3 1

7. Consider, we are given pointers to first and last node of a singly linear linked list, which of the below functions are dependent on the length of the linked list.

   A. Delete the first node from the linked list

   B. Insert a new node as a first node in the linked list

   C. Delete the last node from the linked list

   D. Insert a new node at the end of the linked list

   Answer: (C)

   Explanation:

   A. In order to delete first element length of linked list does not matter.

   B. Similarly, to insert new node at beginning only head is required.

   C. If we want to delete last node of the linked list, first we need to find second last element and last element and for this length of linked list is required. So C is correct option.

   D. In order to add new element at the end, length of linked list does not matter.

8. Consider that a pointer to a node X in a singly linear linked list is given. Pointers to head node is not given, can we delete the node X from given singly linear linked list?

   A. Possible if X is not last node in the linked list. Apply two steps, first copy the data of next of X to X, second delete next of X.

   B. Possible if size of linked list is known prior.

   C. Possible if size of linked list is odd

   D. Possible if X is not first node in the linked list. Apply two steps, first copy the data of next of X to X, second delete next of X.

   Answer: (A)

   Explanation:

   Following are simple steps.

1. Access next node and assign it to temp.

   ```
   struct node *temp  = X->next;
   ```

2. Replace current node(X) with the contents of next node(temp).

   ```
   X->data  = temp->data;
   X->next  = temp->next;
   ```

3. Free next node(temp)

   ```
   free(temp);
   ```

9. Consider the function f defined below.

```
struct item
{
  int data;
  struct item * next;
};

int f(struct item *p)
{
  return (
          (p == NULL) ||
          (p->next == NULL) ||
          (( P->data <= p->next->data) && f(p->next))
         );
}
```

For a given linked list p, the function f returns 1 if and only if (GATE CS 2003)

A. the list is empty or has exactly one element

B. the elements in the list are sorted in non-decreasing order of data value

C. the elements in the list are sorted in non-increasing order of data value

D. not all elements in the list have the same data value.

Answer: (B)
Explanation: The function f() works as follows

1. If linked list is empty return 1
2. Else If linked list has only one element return 1
3. Else if node->data is smaller than equal to node->next->data and again function call with node-->next. if same thing holds for rest of the list then return 1
4. Else return 0

10. Consider, a singly circular linked list is used to represent a Queue. There is only one p variable to access the queue. To which node should p point such that both the operations enQueue and deQueue can be performed in constant time? (GATE 2004)

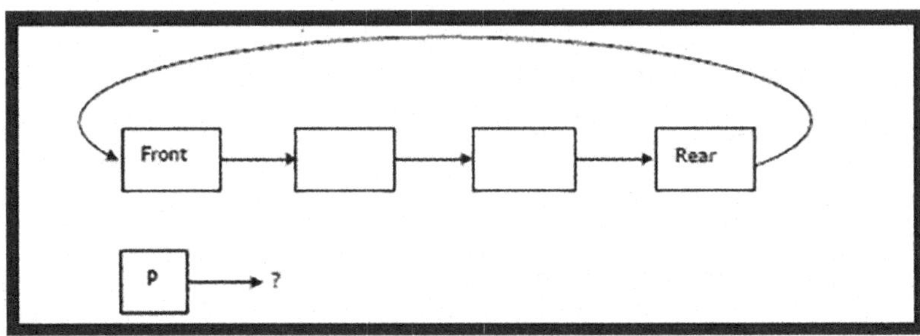

A. rear node
B. front node
C. not possible with a single pointer
D. node next to front

Answer: (A)
Explanation:

A. From rear node we can get front node so able to impement both enQueue and deQueue with the help of rear.
B. From front we cannot access rear so we cannot perform both the operations.
C and D. Not applicable

11. In a circular linked list which of the following is correct,
A. Components are all linked together in some sequential manner.
B. There is no beginning and no end.
C. Components are arranged hierarchically.
D. Forward and backward traversal within the list is permitted.

Answer: (B)

12. In singly linear linked list each node contain minimum of two fields. One field is called as a data field to store the actual data second field is used for to store?
A. Pointer to float data
B. Pointer to integer data
C. Pointer to node
D. Pointer to character data

Answer: (C)

13. Consider the following function that takes reference to head of a Doubly Linked List as parameter. Assume that a node of doubly linked list has previous pointer as *prev* and next pointer as *next*.

```
void fun(struct node **head_ref)
{
    struct node *temp = NULL;
    struct node *current = *head_ref;

    while (current !=  NULL)
    {
        temp = current->prev;
        current->prev = current->next;
        current->next = temp;
        current = current->prev;
    }

    if(temp != NULL )
        *head_ref = temp->prev;
}
```

Assume that reference of head of following doubly linked list is passed to above function
1 2 3 4 5 6.
What should be the modified linked list after the function call?

A. 2 1 4 3 6 5

B. 5 4 3 2 1 6.

C. 6 5 4 3 2 1.

D. 6 5 4 3 1 2

Answer: (C)
Explanation: The given function reverses the given doubly linked list.

14. In the worst case, the number of comparisons needed to search a singly linked list of length n for a given element is (GATE CS 2002)

A. log 2 n

B. n/2

C. log 2 n – 1

D. **n**

Answer: (D)
Explanation: In the worst case, the element to be searched has to be compared with all elements of linked list.

15. A linear collection of data elements where the linear node is given by means of pointer is called?
    A. Linked list
    B. Node list
    C. Primitive list
    D. None

    Answer: (A)

16. Which of the following operations is carried out more efficiently by the doubly-linked list than by the singly-linked list?
    A. Deleting a node whose location in given
    B. Searching of an unsorted list for a given item
    C. Inverting a node after the node with given location
    D. Traversing a list to process each node

    Answer: (D)

17. Consider an implementation of unsorted singly linked list. |||||Suppose it has its representation with a head and tail pointer. Given the representation, which of the following operation can be implemented in O(1) time?
    i. Insertion at the front of the linked list
    ii. Insertion at the end of the linked list
    iii. Deletion of the front node of the linked list
    iv. Deletion of the last node of the linked list
        A. I and II
        B. I and III
        C. I,II and III
        D. I,II and IV

    Answer: (C)

18. Consider an implementation of unsorted singly linked list. Suppose it has its representation with a head pointer only. Given the representation, which of the following operation can be implemented in O(1) time?
    i. Insertion at the front of the linked list
    ii. Insertion at the end of the linked list
    iii. Deletion of the front node of the linked list
    iv. Deletion of the last node of the linked list

A. I and II

B. I and III

C. I, II and III

D. I, II and IV

Answer: (B)

19. Consider an implementation of unsorted circular linked list. Suppose it has its representation with a head pointer only. Given the representation, which of the following operation can be implemented in O(1) time?

 i. Insertion at the front of the linked list

 ii. Insertion at the end of the linked list

 iii. Deletion of the front node of the linked list

 iv. Deletion of the end node of the linked list

 A. I and II

 B. I and III

 C. I, II, III and IV

 D. **None**

Answer: (D)

20. In singly linear linked list each node contains how many minimum fields?

A. Five

B. Zero

C. Two

D. Four

Answer: (C)
Explanation:
In singly linear linked list, each node contains data field and link field.

21. A non-circular doubly linked list can be defined as a......

A. Set of nodes, each with two pointers previous and next

B. Set of nodes chained together with pointers

C. Linear sequence of nodes in sequential memory locations

D. Linear sequence of nodes chained together with pointers

Answer: (D)

22. Which of the below operations is a dictionary operation?
    A. Search an element
    B. Delete an element
    C. Insert an element
    D. All of these

    Answer: (D)

23. What is known as a linear collection of data items, where the each node is given by means of a pointer to the next node?
    A. Linked list
    B. Node list
    C. Primitive list
    D. None of these

    Answer: (A)

24. In singly linear linked list, each node contains minimum of two fields. One field is the data field to store the data second field. What is it called as?
    A. Pointer to character array
    B. Pointer to integer
    C. Node address
    D. Pointer to next linked list node

    Answer: (D)

25. Which of the following type of the linked list can be used so that concatenation of two lists can give the time complexity as O (1)?
    A. Singly linked list
    B. Doubly linked list
    C. Doubly circular linked list
    D. Array implementation of list

    Answer: (C)

26. The linked lists are not appropriate for implementing?
    A. Insertion sort
    B. Radix sort
    C. Polynomial manipulation
    D. Binary search

    Answer: (D)

In the worst case, the number of comparisons needs to search a node in a singly linked list of length n is having time complexity.

A. log n

B. n

C. log2n-1

D. n/2

Answer: (B)

28. Which of the following C language code is used to create a new node in the singly linear linked list.

```
  struct node
{
  int data;
  struct node * next;
}
  typedef struct node LIST;
LIST*ptr;
```

A. **ptr=(LIST*) malloc(sizeof(LIST));**

B. *ptr=(LIST*) malloc(LIST);

C. ptr=(LIST*) malloc(sizeof(LIST*));

D. ptr=(LIST) malloc(sizeof(LIST));

Answer: (A)

29. What is the return type of the malloc () function?

A. void

B. int *

C. float *

D. void *

Answer: (D)
Explanation:
The return type of the malloc () function is void *.

30. LLINK is the pointer toward the....

A. successor node

B. predecessor node

C. head node

D. last node

Answer: (B)

31. Each node of a singly linear linked list must contain at least.....
    A. Three fields
    B. Two fields
    C. Six fields
    D. Five fields

    Answer: (B)

32. Linked list data structures are most appropriate
    A. For comparatively constant collections of data
    B. For the size of the structure and the data within the structure are continuously changing.
    C. Both A and B
    D. None of the above

    Answer: (B)

33. Inserting an element at the start of an array is more difficult than inserting an element at the start of a linked list.
    A. TRUE
    B. FALSE

    Answer: (A)

34. The disadvantage in using a circular linked list is ....
    A. It is possible to enter into an infinite loop
    B. Last node stores the address of first node in circular linked list
    C. Searching is time consuming
    D. Circular linked list requires more memory space

    Answer: (A)

35. Which of the following statement is true?
    a. Using singly linear linked lists and singly circular linked list, it is not possible to traverse the linked list in reverse order.
    b. To find the predecessor of any node, it is required to traverse the list from the first node in case of singly linear linked list.
        A. a only
        B. b only
        C. Both a and b
        D. None of these

    Answer: (C)

36. Which of the following singly linear linked list representation using C language is correct?

A.

```
struct node
{
        int data;
        struct node *next;
};
```

B.

```
struct node
{
        int data;
        struct node **next;
};
```

C.

```
struct node
{
        int data;
        struct node *next;
              struct node *prev;
};
```

D.

```
struct node
{
       int data;
       int node *next;
};
```

Answer: (A)

37. Which of the following statements is false or true?
    1. In doubly linear linked list contains two link fields one is storing the address of next node and the other stores the address of the previous node.
    2. To create a new node dynamically, malloc () function in used.

    A. Statement 1 is false
    B. Statement 2 is false

C. Statements 1 and 2 are false

D. **Statements 1 and 2 are true**

Answer: (D)

38. Which of the following statements is false or true?

1. malloc() function having return type as void * which is generic pointer.
2. malloc () function requires only one argument which is size in interger.

A. Statement 1 is false

B. Statement 2 is false

C. Statements 1 and 2 are false

D. **Statements 1 and 2 are true**

Answer: (D)

39. Which of the following statements is false or true?

1. With linked list, insertion and deletions can be efficiently carried out.
2. Once elements are stored sequentially, it becomes very difficult to insert an element in between or to delete the middle element from the array.

A. Statement 1 is false and Statement 2 is true

B. Statement 2 is false and Statement 1 is true

C. Statements 1 and 2 are false

D. **Statements 1 and 2 are true**

Answer: (D)

40. Which of the following statements is false or true?

1.

```
struct node
{
struct node *prev;
int data;
struct node *next;
}; This is the node representation for Singly Linear
Linked List in C language.
```

2.

```
struct node
```

```
{
     int data;
     struct node *next;
}; This is the node representation for Doubly Linear
Linked List in C language.
```

A. Statement 1 is false and Statement 2 is true

B. Statement 2 is false and Statement 1 is true

C. Statements 1 and 2 are false

D. Statements 1 and 2 are true

Answer: (C)

5

Nonlinear Data Structures: Trees

5.1 Basic Concept and Tree Terminology

5.1.1 Basic Concept

The tree is a nonlinear, non-primitive data structure. Trees are used in many applications to represent the relationship among the data elements that are nodes. Tree is a non-primitive, nonlinear data structure, which organizes data in a hierarchical design. In the tree data structure, every singular element is called a node. Each node in a nonlinear tree data structure stores the actual data of that particular element and the link part store address of the left or right subtree in a hierarchical manner. Tree data structure does not contain any cycle.

In a tree data structure, if we have **N** number of nodes, then we have a maximum of **N-1** number of links or edges.

Example of tree:

5.1.2 Tree Terminology

1. **Root node:**

   In a tree data structure, the first node is called a root node. A parent-free node is referred to as a root node. Every tree needs a root node. The root node may be said to be the origin of the tree structure. In any tree data structure, there should be a single root node. We never have several root nodes within a tree. In Figure 5.1, 'A' is the root node of the tree.

2. **Edge:**

   In a tree data structure, the connecting link between any two nodes is called an edge. A tree with several nodes 'N' will have a maximum number of edges. In Figure 5.1, the link between nodes A and B is referred to as the edge.

3. **Parent:**

   In a tree data structure, the node which is the ancestor of any node is called the parent node. In simple terms, the node which has

DOI: 10.1201/9781003105800-5

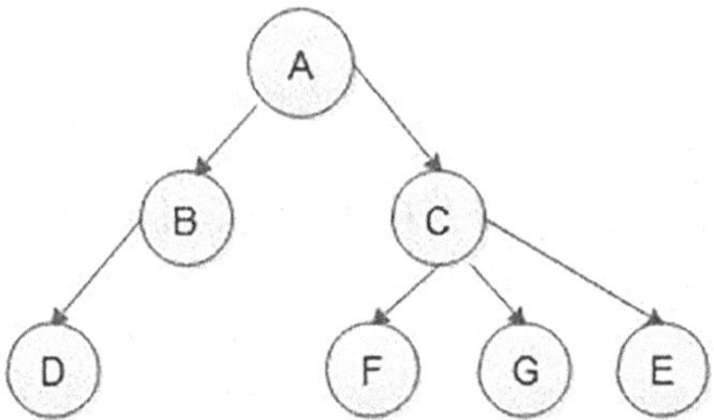

FIGURE 5.1
Tree with seven nodes and six links or edges.

branched from it to any other node is known as the parent node. A parent node can also be described as, "the node that has a child or children". In Figure 5.1. Nodes A, B and C are called parent nodes.

4. **Child:**

   In a tree data structure, the node that is a descendant of any node is referred to as a child node. In simple terms, the node that has an edge on its parent node is called the child node. Within a tree data structure, any parent node may have any number of child nodes. In a tree, all the nodes except the root are child nodes. In Figure 5.1, B, C, D, E, F and G are all child nodes.

5. **Siblings:**

   In a tree data structure, the nodes belonging to the same parent are referred to as brothers and sisters or siblings. In simple words, the nodes with the same parent are called sibling nodes. In Figure 5.1, B and C nodes are siblings of each other. In addition, nodes E, F and G are siblings of each other.

6. **Leaf:**

   In a tree data structure, the node with no child is called a leaf node. In simple terms, a leaf node is a node that has no child.

   In a tree data structure, leaf nodes are also known as external nodes. External node is also a node with no child. In a tree, a leaf node is also called a 'Terminal' node. In Figure 5.1, D, E, F and G are leaf nodes.

7. **Internal nodes:**

   The node that has at least one child is referred to as the internal node. In simple terms, an internal node is a node with at least one child.

   In a tree data structure, nodes other than leaf nodes or external nodes are called internal nodes. The root node is also considered an internal node if the tree structure has more than one node. Internal nodes are also called non-terminal nodes. In Figure 5.1, A, B and C nodes are the internal nodes.

8. **Degree:**

   In a nonlinear tree data structure, the total number of children of a node is known as the degree of that node. Simply stated, the degree of a node in a tree data structure is the total number of child nodes it has.

   The highest degree of a node among all the nodes of a tree is termed as 'degree of tree'. In Figure 5.1, the degree of the C node is 3 and the degree of the B node is 1. However, the degree of tree is 3.

9. **Level:**

   In a nonlinear tree data structure, the root node is stated to be at Level 0 and the children of the root node are at Level 1 and the children of the nodes which are at Level 1 will be at Level 2 and so on. Simply stated, in a tree data structure, each step from top to bottom is called a Level, and the Level count starts with '0' and is incremented by one at each level or step. In Figure 5.1, levels of D, E, F and G are 2.

10. **Height:**

    In a nonlinear tree data structure, the total number of edges or links from a leaf node to a specific node in the longest path is called as height of that node. In a tree, the height of the root node is said to be the height of the tree. In a tree, the height of all leaf nodes is '0'.

    In Figure 5.1, the height of the A node is 2 and the height of tree is also 2.

11. **Depth:**

    In a nonlinear tree data structure, the total number of edges from the root node to a particular node is called the depth of that node. In a tree, the total number of edges from the root node to a leaf node in the longest path is said to be the depth of the tree. Simply stated, the highest depth of any leaf node in a tree is said to be the depth of that tree. In a tree, the depth of the root node is '0'.

    In Figure 5.1, the depth of node D is 2 and the depth of a tree is also 2. The height and depth of a tree are equal, but the height and depth of the node are not equal because the height is calculated by traversing from a given node to the deepest possible leaf node. Depth is calculated from traversing from root to the given node.

12. **Path:**

    In a nonlinear tree data structure, the sequence of nodes and edges from one node to another node is called a path between the two nodes. The length of a path represents the total number of nodes on that path. In Figure 5.1, path A - B - D has length 3.

13. **Sub-tree:**

    In a nonlinear tree data structure, each child of a node forms a subtree recursively. Each child node consists of a sub-tree with its parent node.

14. **Forest:**

    Forest is the disjoint union of trees.

5.2 Data Structure for Binary Trees

5.2.1 Binary Tree Definition with Example

A tree in which each node can have a maximum of two children is called a binary tree. The degree of each node in the binary tree will be at most two (Figure 5.2).

A binary tree with only one node is a binary tree with a root in which the left and right sub-trees are empty. If there are two nodes in a binary tree, one will be the root and another can be either the left or the right child of the root (Figure 5.3).

5.2.2 Types of Binary Tree

1. **Strictly binary tree or full binary tree or proper binary tree:**
   A full binary tree or a proper binary tree or a strictly binary tree is a tree in which each node other than the leaves has two children. This means each node in the binary tree will have either 0 or 2 children is said to be a full binary tree or strictly binary tree. A complete binary tree is used to depict mathematical expressions (Figure 5.4).

2. **Complete binary tree or perfect binary tree:**
   A complete binary tree is a binary tree in which every level, except possibly the last, is filled, and all nodes are as far left as possible. All complete binary trees are strictly binary trees, but all strictly binary trees are not complete binary trees. A complete binary tree

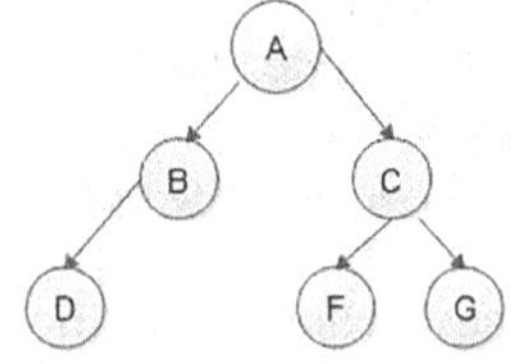

FIGURE 5.2
Binary tree.

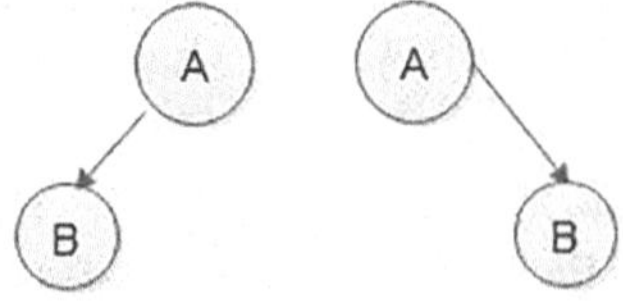

FIGURE 5.3
Two binary trees with two nodes.

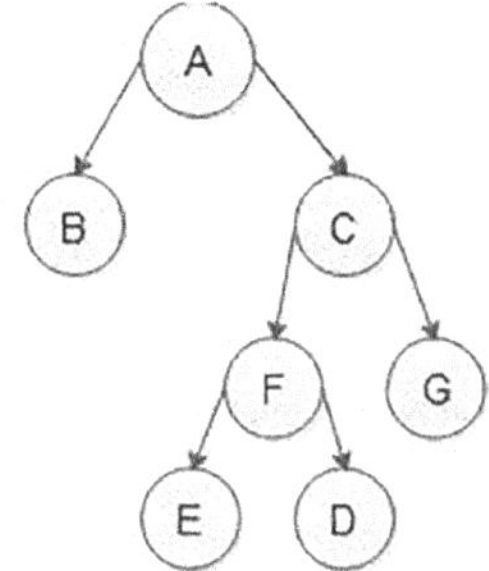

FIGURE 5.4
Strictly binary tree or **full binary tree** or **proper binary tree**.

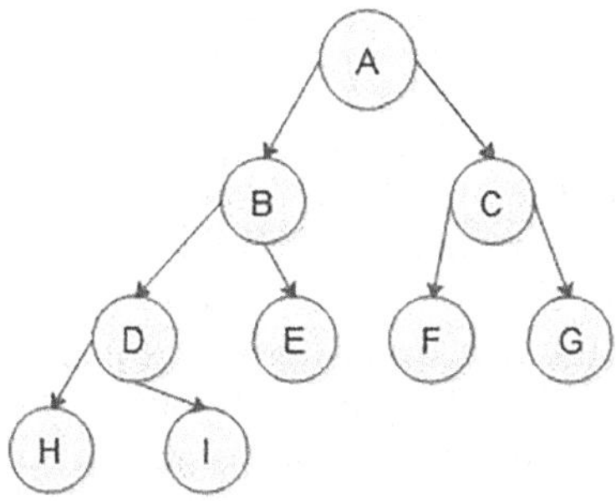

FIGURE 5.5
Complete binary tree.

is the first strictly binary tree with the height difference of only 1 allowed between the left and right sub-tree. All leaf nodes are either at the same level or only a difference of 1 level is permissible (Figure 5.5).

3. **Binary search tree:**

   If a binary tree has the property that all elements in the left sub-tree of a node n are lesser than the contents of n and all elements in the right sub-tree are greater than or equal to the contents of n, such a binary tree is called a binary search tree (BST).

   In a BST, if the same elements are inserted, then the same element is inserted on the right side of the node, having the same value. A BST is shown in Figure 5.6.

4. **Almost complete binary tree:**

   A binary tree of depth d is called an almost complete binary tree if and only if:

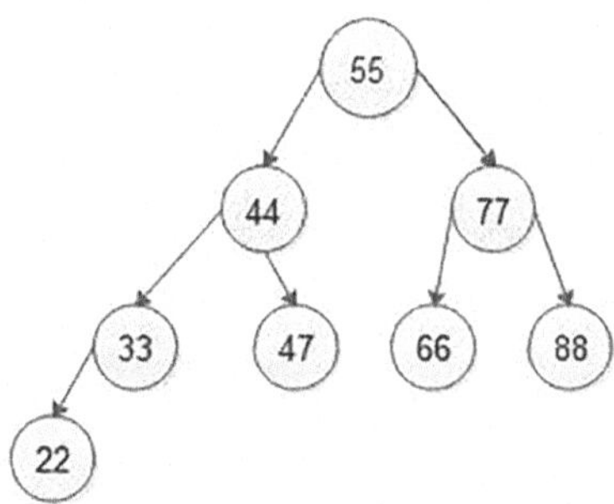

FIGURE 5.6
Binary search tree.

1. Each leaf node in the tree is either at level d or at level d-1, which means the difference between levels of leaf nodes is only one is permissible.
2. At depth d, that is the last level, if a node is present, then all the nodes to the left of that node should also be present first, and no gap between the two nodes from the left to the right side of level d.

   In Figure 5.7, nodes at the last level are filled from left to right, and no gaps between the last level's nodes. However, in Figure 8, the last level, having two nodes H and I have one gap because D nodes left child is present, but the right child is absent so Figure 5.8 is not an almost complete binary tree.

   Figure 5.7 shows almost a complete binary tree, but not a complete binary tree because the D node has a left child but no right child. In the complete binary tree, every interior node has exactly two children.

5. **Skewed binary tree:**

   If a tree is dominated by the left child node or the right child node, it is said to be a skewed binary tree.

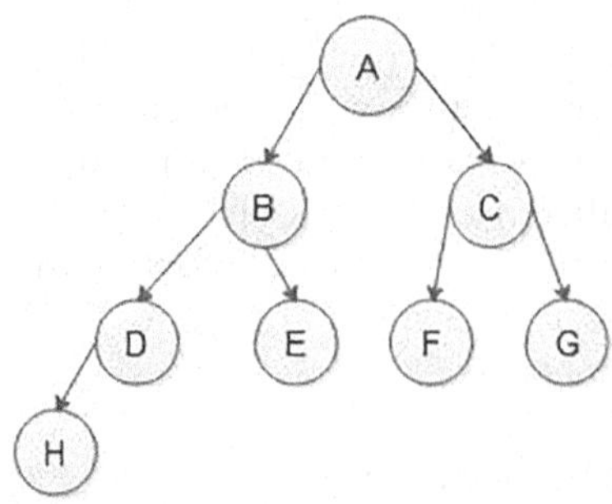

FIGURE 5.7
Almost complete binary tree.

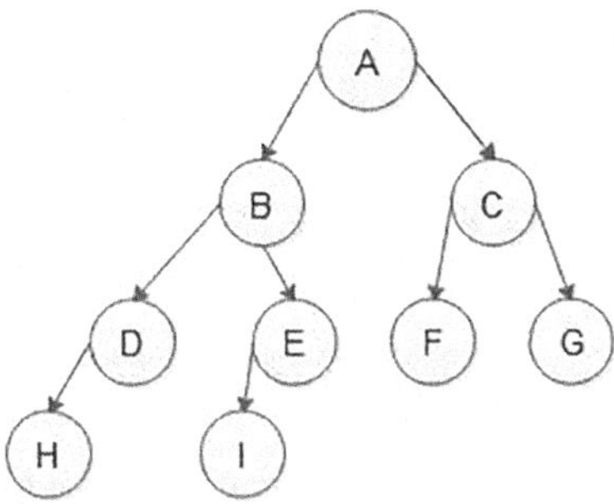

FIGURE 5.8
Not an almost complete binary tree.

In a left-skewed tree, most of the nodes have a left child without a corresponding right child as shown in Figure 5.9. In a right-skewed tree, most of the nodes have the right child without a corresponding left child as shown in Figure 5.10.

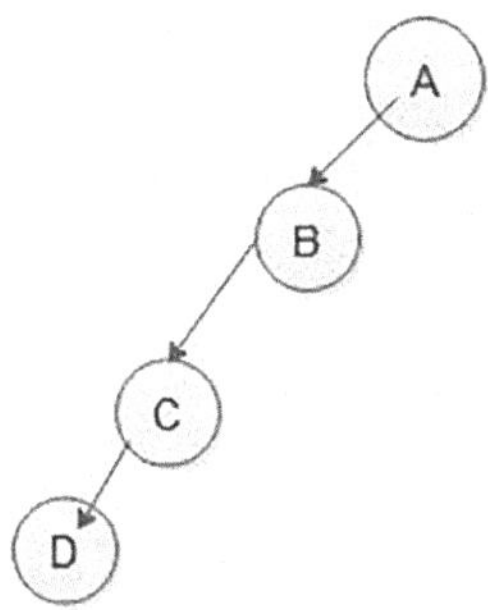

FIGURE 5.9
Left skewed tree.

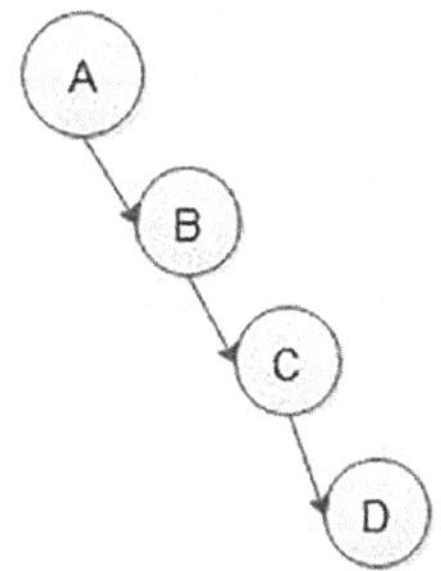

FIGURE 5.10
Right skewed tree.

5.2.3 Binary Tree Representation

There are two binary tree representation techniques:

1. Sequential representation or array representation
2. Linked representation or linked list representation

5.2.3.1 Sequential Representation or Array Representation

If a binary tree is a complete binary tree and if the depth of a tree is 3, then the total number of nodes is 7. That is, if the complete binary tree has depth n, then max nodes are calculated by $2n - 1$ (Figure 5.11).

If the depth of a binary tree is known, then we can represent binary trees using an array data structure. In sequential representation, the numbering of each node will start from the root node, and then the remaining nodes will give ever-increasing numbers in a level-wise direction. The nodes on the same level will be numbered from left to right. The numbering will be as shown in Figure 5.12.

If we know the index of parent node suppose n, then its left child node is represented as,

Left child (n) = (2n+1)

If we know the index of parent node suppose n, then its right child node is represented as,

Right child (n) = (2n+2)

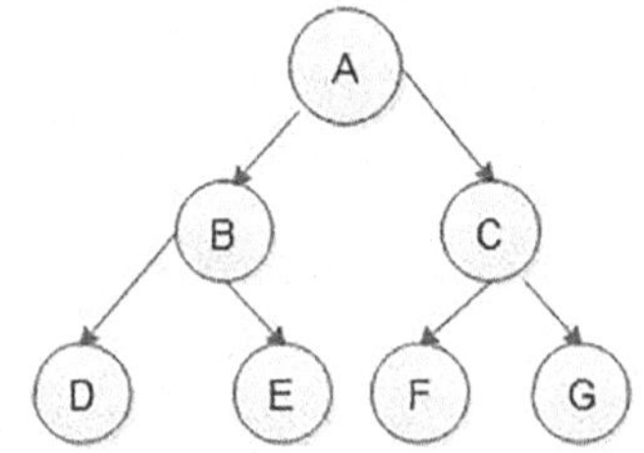

FIGURE 5.11
Complete binary tree.

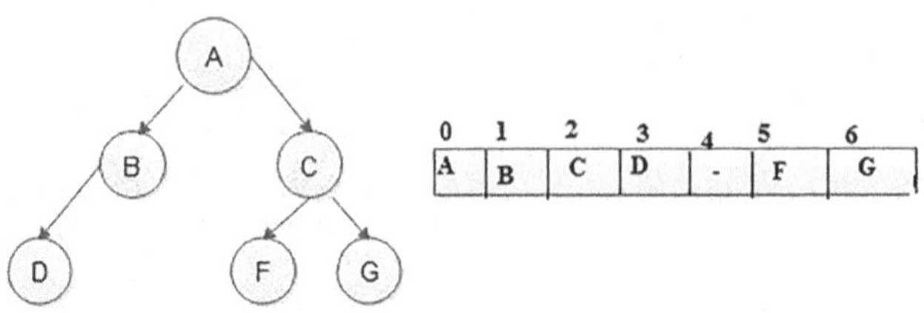

FIGURE 5.12
Sequential representation of binary tree using array.

Here, root node is at index 0.

In Figure 5.12, node C is at index 2, then its left child is at index 5 and the right child is at index 6. In addition, if we know the index of the left child, then we can calculate the index of its parent node using formula (n-1)/2, and if we know the index of the right child, then we can calculate the index of its parent node using formula (n-2)/2.

Advantages of sequential representation:

1. Direct access to any node can be possible.
2. Finding parent or finding left, the right child of any particular node is faster.

Disadvantages of sequential representation:

1. Wastage of memory: If a binary tree is not a complete binary tree, then array size increased. For example, in Figure 5.12, the depth of a tree is 3, then the maximum size of the array is $2^3 - 1$, that is, 7, but actual array utilization is 6.
2. In this type of representation, prior knowledge of the depth of a tree is required because on that basis we have to fix the size of the array.
3. Insertion and deletion of any node in the tree will be costlier as other nodes have to be adjusted at appropriate positions. Therefore, the meaning of a binary tree can be preserved. As there are the above-mentioned drawbacks in the sequential representation, we will search for a more flexible representation. Therefore, instead of the array, we will make use of linked list representation to represent a binary tree.

5.2.3.2 Linked List Representation

In linked list representation, each node will look like this (Figure 5.13):

Left Child	Data	Right Child

The C structure for nodes in the binary tree is as follows:

```
struct binary_tree
{
struct bt *left;
int data;
struct bt *right;
};
```

Left Child	Data	Right Child

FIGURE 5.13
Node in the linked list representation.

Advantages of Linked list representation:

1. There is no wastage of memory.
2. There is no need to have prior knowledge of the depth of the tree. Using the dynamic memory management concept, we can create as much memory or nodes as required.
3. Insertion and deletion of nodes that are the most common operations can be done without moving other nodes.

Disadvantages of Linked list representation:

1. This representation gives no direct access to a node, and special algorithms are required.
2. This representation needs additional space in each node to store left and right sub-tree addresses.
3. The direct address of the left and right child does not get in linked representation.

5.2.4 Operations on Binary Tree

1. To create a binary tree.
2. To display data in each node of binary tree that is tree traversal.
3. To insert any node in the binary tree as the left child or right child of any node.
4. To delete a node from a binary tree.

5.2.5 Binary Tree Traversals

Traversing the tree means visiting each node in the tree exactly once. There are six ways to traverse a tree. For these traversals, we will use some notations as:

L means move to left child.

R means move to right child.

R′ means move to root or parent node.

Therefore, there are six different combinations of L, R, R′ nodes as LR′R, LRR′, R′LR, R′RL, RR′L and RLR′.

However, from there three traversing methods are important, which are as follows:

1. Left-Root-Right (LR'R): In-order traversal.
2. Root-Left-Right (R'LR): Preorder traversal.
3. Left-Right-Root (R'LR): Post-order traversal.

These all above traversing technique is also called as **Depth First searching (DFS) or Depth First Traversal.**

A tree is typically traversed in two ways:

1. Breadth first traversal (BFS) or level order traversal
2. Depth first traversal (DFS)

5.2.5.1 Breadth First Traversal or Level Order Traversal

In the above traverse method, the items are visited level-wise. Therefore, the first node A is visited, then at the second level, first the nodes B then C are visited, and then at the third level, first nodes D, then F and then G are visited (Figure 5.14).

Level Order Traversal is as follows: A - B - C - D - F – G

Algorithm

In a binary tree, the root node is visited first, then its child nodes are visited and enqueue in a queue. Then dequeue a node from the queue and find its child nodes and again insert it into queue till all nodes are visited.

Algorithm for Level Order Traversal in Binary tree

1. Create an empty queue q.
2. temp_node = root **// Start from the root node and assign the value of root to temp_node**

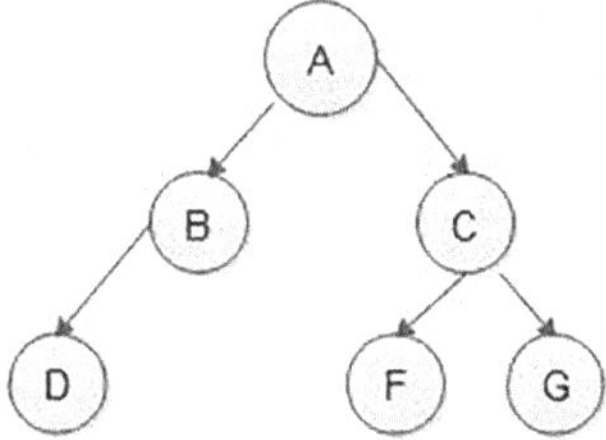

FIGURE 5.14
Binary tree traverse by using **Breadth First Traversal or Level Order Traversal.**

3. while temp_node is not NULL repeat the steps
   3.1. Print temp_node->data.
   3.2. Enqueue temp_node's children from first left then right children to q.
   3.3. Dequeue a node from q and assign its value to temp_node.
4. Stop.

Time Complexity of Breadth First Traversal or Level Order Traversal: O (n) because we visit every node exactly once.

Extra space required for level order traversal is O (w) where w is the maximum width of a binary tree. In breadth first traversal, a queue data structure is required. In breadth first traversal, queue data structure stores one by one node of different levels.

5.2.5.2 Depth First Traversal

Depth first traversal has following three types of traversing:

1. In-order traversal
2. Preorder traversal
3. Post-order traversal

5.2.5.2.1 In-order Traversal

The basic principle is to traverse the left sub-tree, then root and then the right sub-tree (Figure 5.15).

In-order traversal is as follows: D - B - A - F - C – G

5.2.5.2.2 Preorder Traversal

The basic principle is to traverse the first root, then the left sub-tree and then the right sub-tree (Figure 5.16).

Preorder traversal is as follows: A - B - D - C - F – G

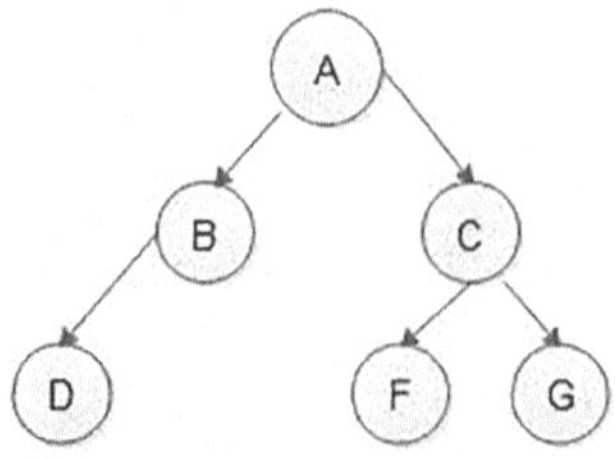

FIGURE 5.15
Binary tree traverse by using in-order traversal method.

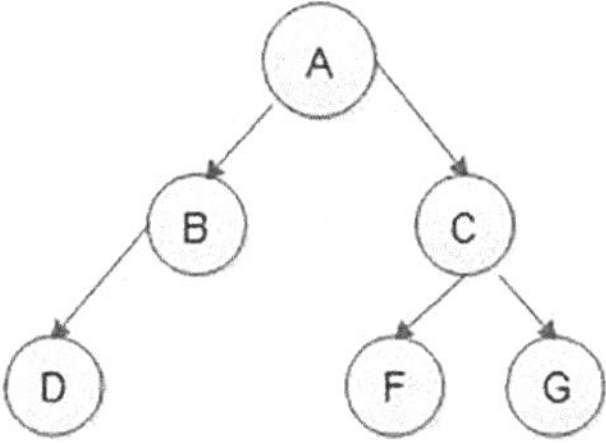

FIGURE 5.16
Binary tree traverse by using preorder traversal method.

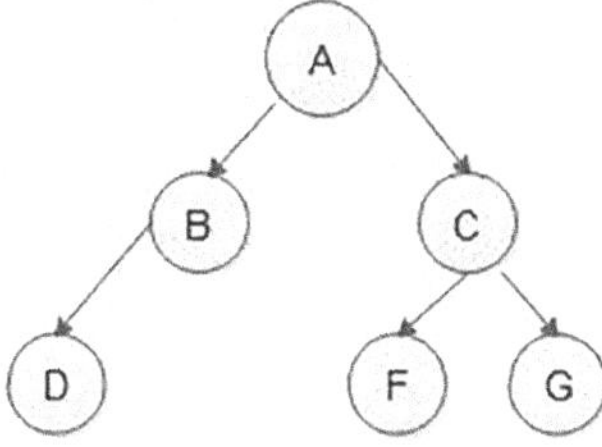

FIGURE 5.17
Binary tree traverse by using post-order traversal method.

5.2.5.2.3 Post-order Traversal

The basic principle is to traverse the first left sub-tree, then the right sub-tree and then root (Figure 5.17).

Post-order traversal is as follows: D - B - F - G - C – A

5.2.6 Simple Binary Tree Implementation

Creating a simple binary tree, the first element will form the root node. For all the next elements, the user has to enter whether that element has to be inserted as a left child or a right child of the previous node.

Example:

Let us say 10 is the root node,

If the next node says 21, then users will be asked for his choice that is attached to the left or right of 10. If the user answers left or l, then the tree will be,

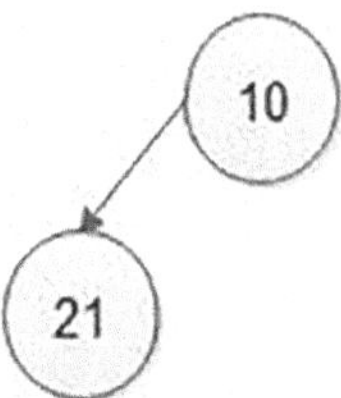

Then, if the next element is 11, then again the user has to give his choice, whether it is left or right. That means first whether the user wants to attach 11 left or right of root 10. Then, if the user answers left, then again the user asks whether the user wants to attach 11 to left or right of 21, and if the user answers right, then node 11 will be attached as a right child of 21.

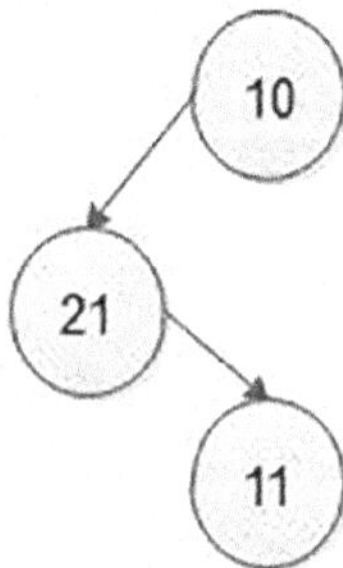

Thus, a simple binary will be generated. The principal idea behind this creation is to always ask the user where he wants to attach the next node and always start scanning the tree from the root.

Program: Implementation of simple binary tree and performing recursive insertion and traversing operations on a simple binary tree.

```
#include<stdio.h>
#include<stdlib.h>
struct bt
{
struct bt *left;
int data;
struct bt *right;
};
// Function prototypes
void insert(struct bt *root, struct bt *new1);
void in-order(struct bt *temp);
void preorder(struct bt *temp);
void postorder(struct bt *temp);
// main function definition
```

```
int main()
{
struct bt *root, *new1;
int ch;
char c;
root=NULL;
do
{
printf("1. Create 2.In-order 3. Preorder 4. Postorder 5. Exit
\n");
printf("Enter your choice: ");
scanf("%d",&ch);
switch(ch)
{
case 1:
do
       {
       new1=(struct bt*)malloc(sizeof(struct bt));
       printf("\n Enter data: ");
       scanf("%d",&new1->data);
       new1->left=NULL;
       new1->right=NULL;
       if(root==NULL)
       {
       root=new1;
       }
       else
       {
       insert(root,new1);
       }
       printf("\n Do you want to insert new node [y/n]: ");
       c = getche();
       }while(c=='y'||c=='Y');
       break;
case 2:
       if(root==NULL)
       printf("\n Binary Tree is not created");
       else
       {
       in-order(root);
       }
       break;
case 3:
       if(root==NULL)
       printf("Binary Tree is not created");
       else
       {
```

```
        preorder(root);
        }
        break;
case 4:
        if(root==NULL)
        printf("Binary Tree is not created");
        else
        {
        postorder(root);
        }
        break;
case 5:
       exit(0);
default:
       printf("\n wrong choice: ");
}
printf("\n Do you want to continue Binary Tree [y/n] : ");
fflush(stdin);
scanf("%c",&c);
}while(c=='y'||c=='Y');
return 0;
}
void insert(struct bt* root, struct bt *new1)
{
char c;
printf("\n Where to insert node right or left of %d  : ",
root->data);
c=getche();
//if new node is to be inserted at right sub-tree
if(c=='r'||c=='R')
{
    if(root->right==NULL)  //if root node's right pointer is
    NULL then insert new node to
                                                   // right side
    {
    root->right=new1;
    }
    else                 // otherwise call insert function
    recursively
    {
    insert(root->right,new1);
}
}
//if new node is to be inserted at left sub-tree
if(c=='l'||c=='L')
{
if(root->left==NULL)
```

```
{
root->left=new1;
}
else
{
insert(root->left,new1);
}
}
}
//in-order traversing of a simple binary tree
void in-order(struct bt *temp) //pass root node address as an
argument to a in-order function
{
if(temp != NULL)
{
in-order(temp->left); // Traverse the left sub-tree of root R
printf("%d\t", temp->data);  //Visit the root, R and display.
in-order(temp->right);  //Traverse the right sub-tree of root
R
}
}
//preorder traversing of a simple binary tree
void preorder(struct bt *temp) //pass root node address as an
argument to a preorder
                                                        //
      function
{
if(temp != NULL)
{
printf("%d\t", temp->data); //Visit the root, R and display.
preorder(temp->left); // Traverse the left sub-tree of root R
preorder(temp->right); //Traverse the right sub-tree of root R
}
}
//post-order traversing of a simple binary tree
void postorder(struct bt *temp) //pass root node address as an
argument to post-order
                                    //function
{
if(temp!=NULL)
{
postorder(temp->left);  // Traverse the left sub-tree of root
R
postorder(temp->right); //Traverse the right sub-tree of root
R
printf("%d\t",temp->data); //Visit the root, R and display.
}
}
```

Output:

```
1. Create 2.Inorder 3. Preorder 4. Postorder 5. Exit
Enter your choice: 1

 Enter data: 21

 Do you want to insert new node [y/n]: y
 Enter data: 31

 Where to insert node right or left of 21  : l
 Do you want to insert new node [y/n]: y
 Enter data: 41

 Where to insert node right or left of 21  : r
 Do you want to insert new node [y/n]: y
 Enter data: 51

 Where to insert node right or left of 21  : r
 Where to insert node right or left of 41  : r
 Do you want to insert new node [y/n]: y
 Enter data: 61

 Where to insert node right or left of 21  : l
 Where to insert node right or left of 31  : l
 Do you want to insert new node [y/n]: y
 Enter data: 71

 Where to insert node right or left of 21  : l
 Where to insert node right or left of 31  : r
 Do you want to insert new node [y/n]: n
 Do you want to continue Binary Tree [y/n] : y
1. Create 2.Inorder 3. Preorder 4. Postorder 5. Exit
Enter your choice: 2
61       31       71       21       41       51
 Do you want to continue Binary Tree [y/n] : y
1. Create 2.Inorder 3. Preorder 4. Postorder 5. Exit
Enter your choice: 4
61       71       31       51       41       21
 Do you want to continue Binary Tree [y/n] : n
```

5.3 Algorithms for Tree Traversals

5.3.1 Recursive Algorithm for In-order Traversal

In-order traversing of a binary tree T with root R.

Algorithm in-order (root):

1. Traverse the left sub-tree of root R, that is, call in-order (left sub-tree).
2. Visit the root, R and process.
3. Traverse the right sub-tree of root R, that is, call in-order (right sub-tree).

5.3.2 Recursive Algorithm for Preorder Traversal

Preorder traversing of binary tree T with root R.

Algorithm preorder (root):

1. Visit the root, R and process.
2. Traverse the left sub-tree of root R, that is, call preorder (left sub-tree)
3. Traverse the right sub-tree of root R, that is, call preorder (right sub-tree)

5.3.3 Recursive Algorithm for Postorder Traversal

Postorder traversing of binary tree T with root R.

Algorithm Postorder (Root)

1. Traverse the left sub-tree of root R, that is, call postorder (left sub-tree)
2. Traverse the right sub-tree of root R, that is, call postorder (right sub-tree)
3. Visit the root, R and process.

5.4 Construct a Binary Tree from Given Traversing Methods

5.4.1 Construct a Binary Tree from Given In-order and Preorder Traversing

Take one example:

In-order traversal: 14, 7, 21, 8, 15, 11, 61, 99 and 33

Preorder traversal: 11, 21, 14, 7, 15, 8, 33, 61 and 99

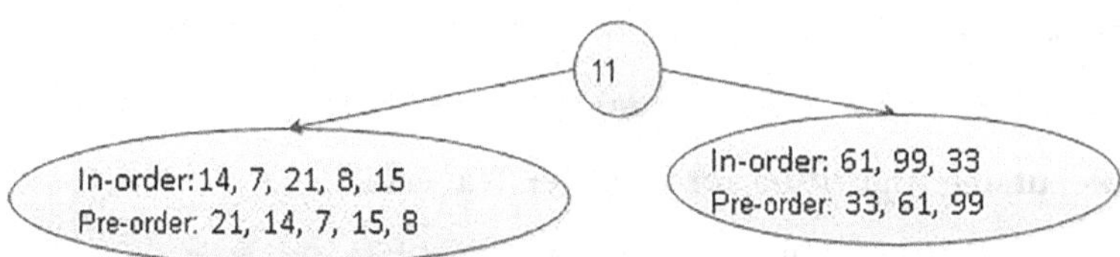

FIGURE 5.18
Reconstruction of a binary tree step 1.

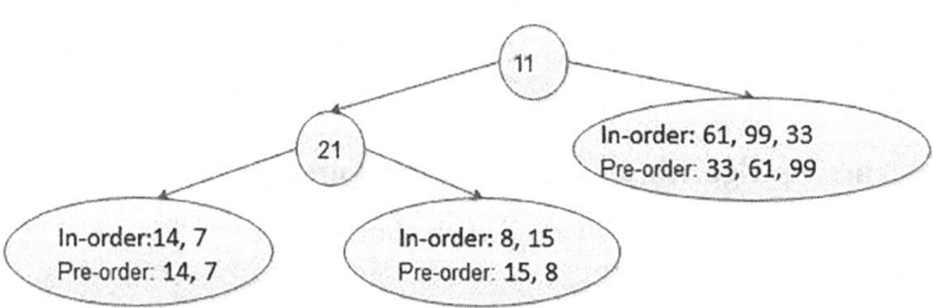

FIGURE 5.19
Reconstruction of a binary tree step 2.

The first value in the preorder traversal is the root of the binary tree because in preorder traversing, the first root is traversed, then the left sub-tree and then the right sub-tree is traversed. Therefore, the node with data 11 becomes the root of the binary tree. In in-order traversal, initially, the left sub-tree is traversed, then the root node and then the right sub-tree. Therefore, the data before 11 in the in-order list, that is, 14, 7, 21, 8 and 15 forms the left sub-tree, and the data after 11 in the in-order list, that is, 61, 99 and 33 forms the right sub-tree. Figure 5.18 shows the structure of the tree after dividing the tree into left and right sub-trees.

In the next iteration, 21 is the root for the left sub-tree. Hence, the data before 21 in the in-order list, that is, 14, 7 form the left sub-tree of the node that contains a value 21. The data that come to the right of 21 in the in-order list that is 8, 15 forms the right sub-tree of the node with value 21. Figure 5.19 shows the structure of the tree after expanding the left and right sub-tree of the node that contains a value 21.

Now the next data in the preorder list is 14 so now it becomes the root node. The data before 14 in the in-order list forms the left sub-tree of the node that contains a value 14. However, as there is no data present before 14 in in-order list, the left sub-tree of the node with value 14 is empty. The data that come to the right of 14 in the in-order list is 7 forms the right sub-tree of the node that contains a value 14. Figure 5.20 shows the structure of the tree after expanding the left and right sub-tree of the node that contains a value 14.

In the same way, one by one all the data are picked from the preorder list and are treated as root and then from in-order list find elements in the right and left sub-tree, and the whole tree is constructed in a recursive way or non-recursive way, that is, iterative way (Figures 5.21 and 5.22).

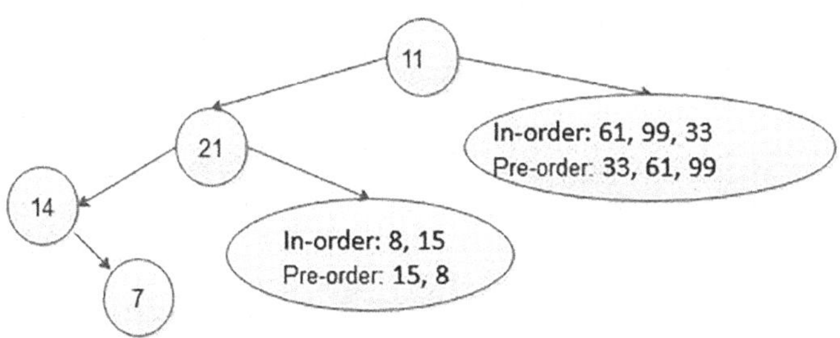

FIGURE 5.20
Reconstruction of a binary tree step 3.

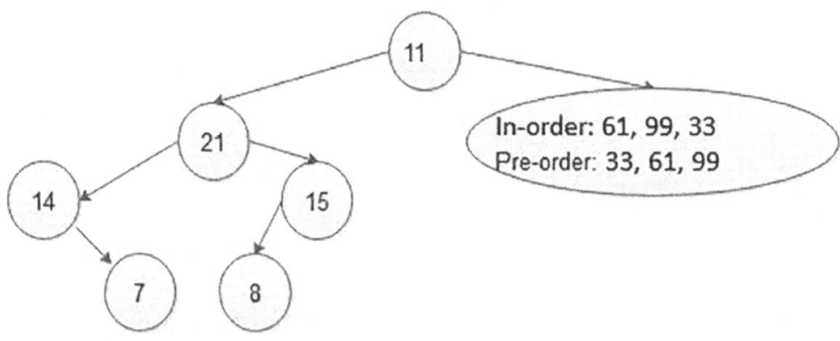

FIGURE 5.21
Reconstruction of a binary tree step 4.

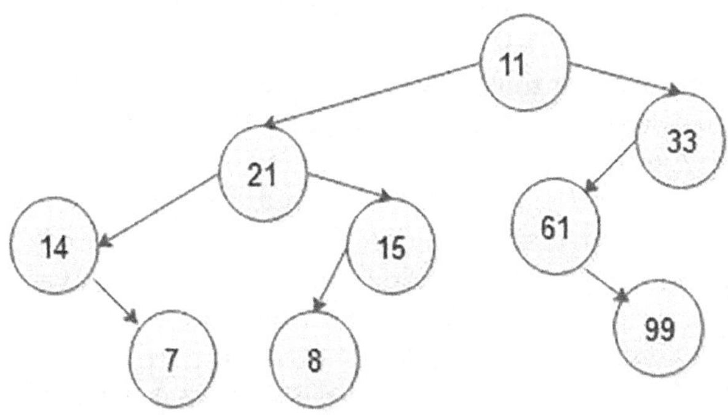

FIGURE 5.22
Reconstruction of a binary tree step 5.

5.4.2 Construct a Binary Tree from Given In-order and Post-order Traversing

Take an example:

In-order traversal: 4 2 5 1 6 7 3 8

Post-order traversal: 4 5 2 6 7 8 3 1

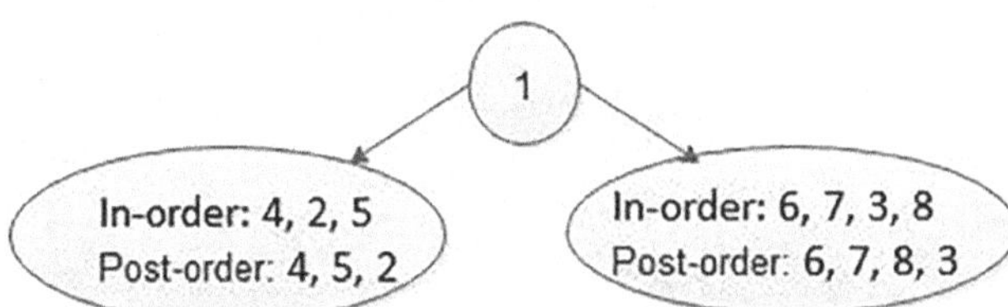

FIGURE 5.23
Reconstruction of a binary tree step 1.

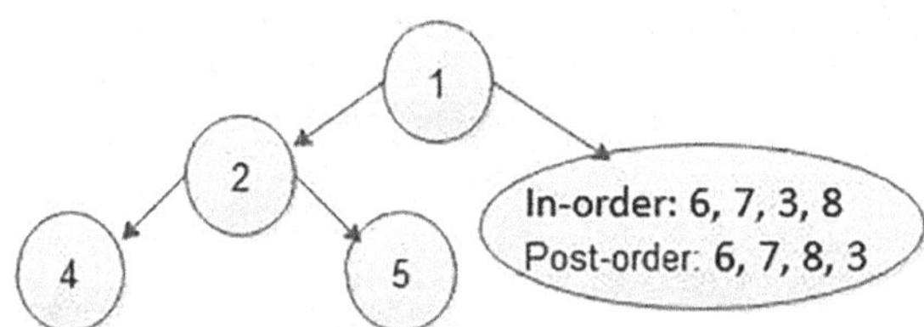

FIGURE 5.24
Reconstruction of a binary tree step 2.

The last value in the post-order traversal is the root of the binary tree because in post-order traversing root is traversed in the last then the left sub-tree and then the right sub-tree is traversed. Therefore, the node with data 1 becomes the root of the binary tree. In in-order traversal, initially the left sub-tree is traversed, then the root node and then the right sub-tree. Therefore, the data before 1 in the in-order list that is 4, 2, 5 forms the left sub-tree and the data after 1 in the in-order list that is 6, 7, 3, 8 forms the right sub-tree. Figure 5.23, shows the structure of the tree after dividing the tree into the left and right sub-trees.

In the next iteration 2 is root for the left sub-tree because it is the last element in post-order traversing. Hence, the data before 2 in the in-order list that is 4 forms the left sub-tree of the node that contains a value 2. The data that comes to the right of 2 in the in-order list that is 5 forms the right sub-tree of the node with value 2. Figure 5.24 shows the structure of the tree after expanding the left and right sub-tree of the node that contains a value 2.

In the same way, one by one last data elements are picked from the post-order list and are treated as root, and then from the in-order list find elements in the right and left sub-tree and the whole tree is constructed in a recursive way or non-recursive way (Figures 5.25 and 5.26).

5.4.3 Construct a Strictly or Full Binary Tree from Given Preorder and Post-order Traversing

If we have preorder and post-order traversing sequence, then we cannot form or construct simple or general binary tree. Because for two different binary

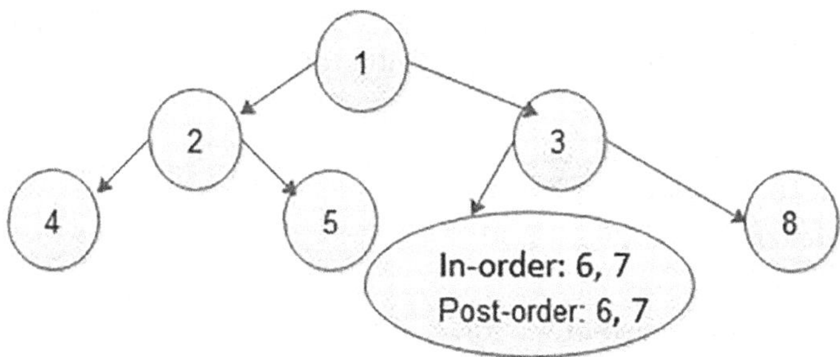

FIGURE 5.25
Reconstruction of a binary tree step 3.

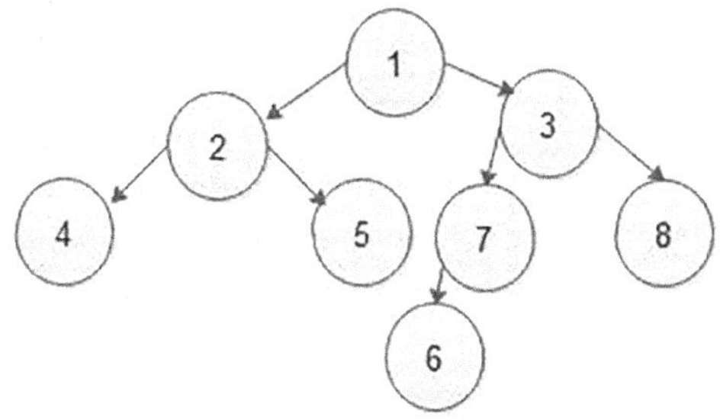

FIGURE 5.26
Reconstruction of a binary tree step 4.

trees post and preorder traversing the sequence may be the same as shown in Figure 5.27.

Preorder traversal: 1 - 2 - 3

Post-order traversal: 3 - 2 - 1

In Figure 5.27a and b, both preorder and post-order traversal give us the same result. That is, using preorder and post-order traversal, we cannot predict exactly which binary tree we want to construct. That means ambiguity is there.

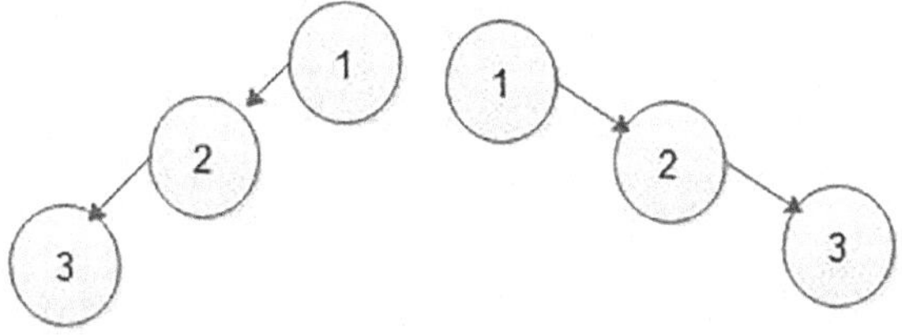

FIGURE 5.27
Binary tree.

However, if we know that the binary tree is strictly or a full binary tree, we can construct the binary tree without ambiguity.

Take an example:

Preorder traversal: 1 2 4 5 3 7 8

Post-order traversal: 4 5 2 7 8 3 1

The last value in the post-order traversal is the root of the binary tree. The first value in preorder traversal is also root. Therefore, the node with data 1 becomes the root of the binary tree. The value next to 1 in preorder, must be left child of the root. Therefore, we know 1 is root and 2 is the left child. Now we know 2 is the root of all nodes in the left sub-tree. All nodes before 2 in post-order must be in the left sub-tree that is 4, 5 and 2. Now we know 1 is the root, elements 4, 5 and 2 are in the left sub-tree, and the elements 7, 8 and 3 are in the right sub-tree (Figures 5.28 and 5.29).

In the same way, one by one last data elements are picked from the post-order list and are treated as root and then from in-order list find elements in the right and left sub-tree, and the whole tree is constructed either in a recursive way or in a non-recursive way, i.e., iterative way (Figure 5.30).

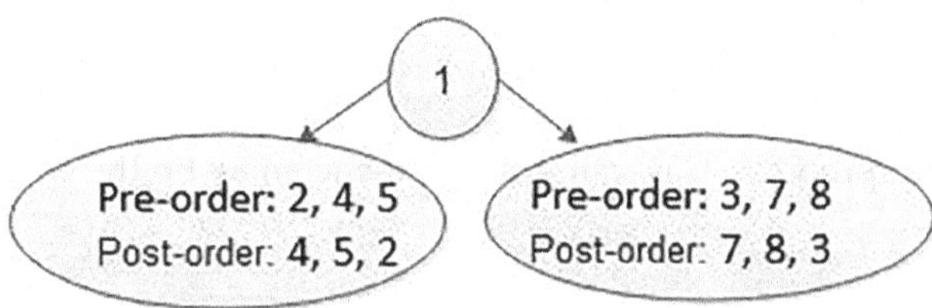

FIGURE 5.28
Reconstruction of a full binary tree step 1.

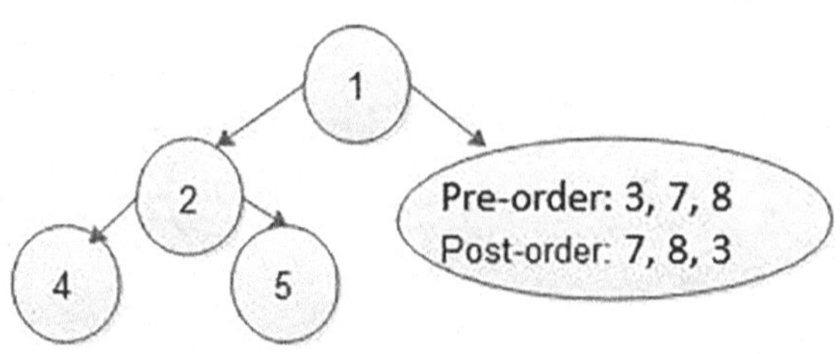

FIGURE 5.29
Reconstruction of a full binary tree step 2.

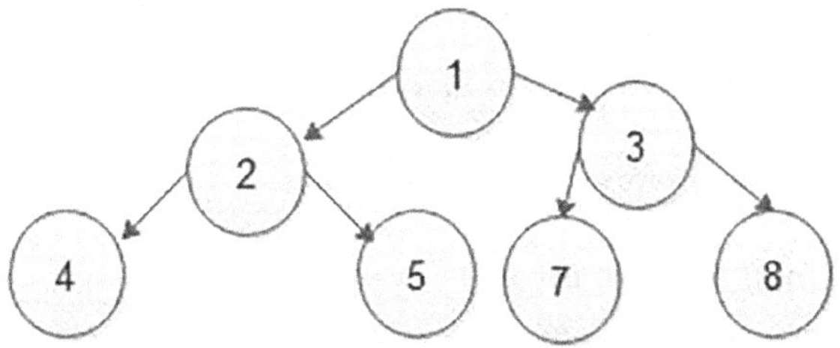

FIGURE 5.30
Reconstruction of a full binary tree step 3.

5.5 Binary Search Trees

5.5.1 Basic Concepts of BSTs

In a simple binary tree, nodes are located in any fashion. According to the user's aspiration, new nodes can be connected as a left or right child of any desired node. In such a case, finding for any node is a long cut procedure because we have to search the entire tree. In addition, the complexity of the search time is going to increase unnecessarily. Therefore, to make the search algorithm quicker in a binary tree, let us create the BST. The BST is derived from a binary search algorithm. During the creation of a BST, the data are systematically organized. That means, the value of the left sub-tree < root node < value at the right sub-tree (Figure 5.31).

5.5.2 Operations on BSTs

1. Insertion of nodes in the BST

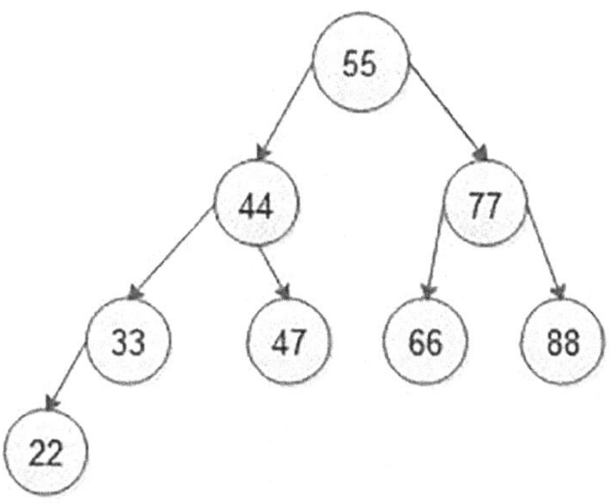

FIGURE 5.31
Binary search trees.

2. Deletion of a node from BST
3. Searching for a particular node in BST

Let us discuss these operations with examples:

5.5.2.1 Insertion of Node in the BST

While inserting any node in BST, first of all, we have to look for its appropriate position in BST.

In Figure 5.32, if we want to insert 32, then we will start comparing 32 with a value of root, that is, 55. As 32 is less than 55, we will move on the left sub-tree. Now we will compare 32 with 15, and 32 is greater than 15 so move on the right sub-tree. Again, compare 32 with 33, and 32 is less than 33 so move to left sub-tree, but the left sub-tree is empty, so attach 32 nodes on the left side of 33 nodes as the left child of 33 nodes (Figure 5.33).

5.5.2.2 Deletion of Node in the BST

For deleting a node from a BST, there are three possible cases:

1. Deletion of the leaf node from a BST
2. Deletion of a node having one child as left or right
3. Deletion of a node having two children.

5.5.2.2.1 Deletion of a Leaf Node

This is the simplest deletion, in which we first search the node which we want to delete and its respective parent. If the node which we want to delete is a leaf node, then that leaf node is either left or right child of its respective parent, then set the left or right pointer of the parent node as NULL (Figure 5.34) (Figure 5.34).

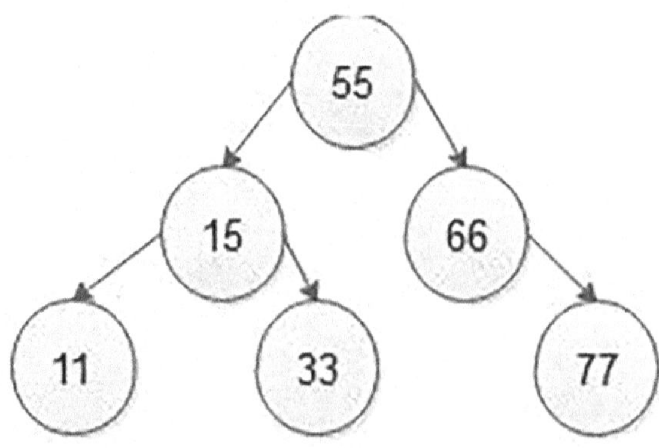

FIGURE 5.32
Before insertion of a node in the binary search tree.

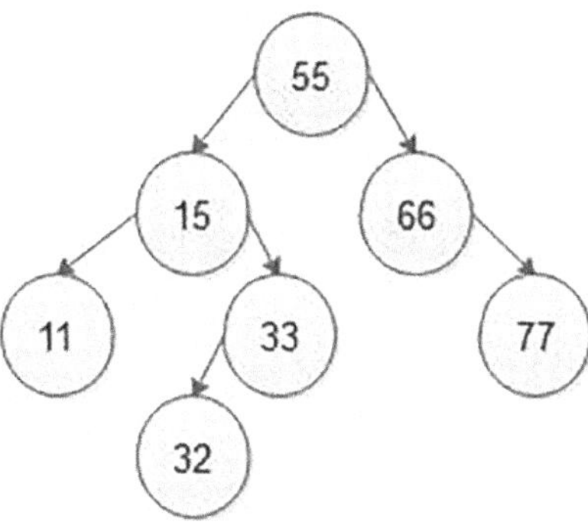

FIGURE 5.33
After insertion of node 32 in a binary search tree.

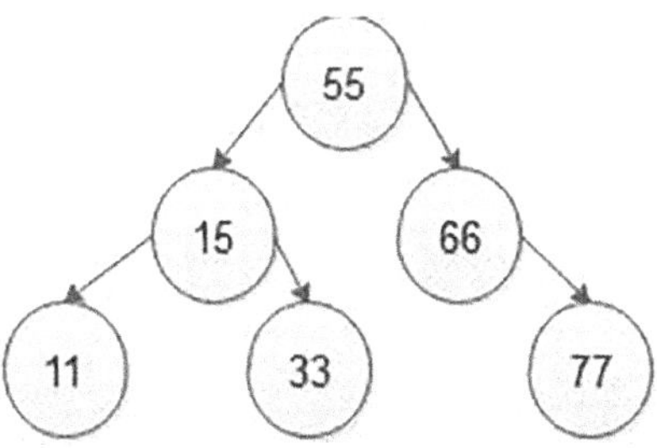

FIGURE 5.34
Before deletion of a leaf node from binary search tree.

In the above BST, the node having value 33 is a leaf node, that node we want to delete means that we will first search the node 33 in the BST and find its parent node that is 15. Then find 33 nodes are left or right side of the 15-parent node. In the above example, 33 nodes are on the right side of parent node 15 so make 15 node's right pointer as NULL. In addition, free the node 33 as is shown in Figure 5.35.

5.5.2.2.2 Deletion of a Node Having One Child as Left or Right

If the node which we want to delete from the BST is an interior node having one right child, then find its parent node and check whether that deleted node is a left or right child of the parent node. If the deleted node is the right child of the parent node, then copy the address of the right child of the deleted node into the right side of the parent node (Figure 5.36).

We want to delete node 33 in the above BST because it has an only left child. To do so, we will first search node 33 in the BST and find its parent node, that is, 15. Then find if the 33 node is on the left or right side of the 15 parent node. In the above example, 33 node is at the right side of parent node 15 so copy the address of the left child of deleted node 33 to the right side of parent node 15. Then set deleted node as free from BST as is shown in Figure 5.37.

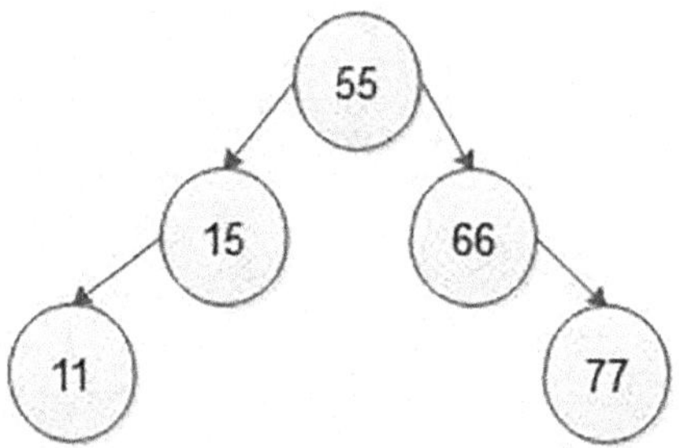

FIGURE 5.35
After deletion of node 33 from binary search tree.

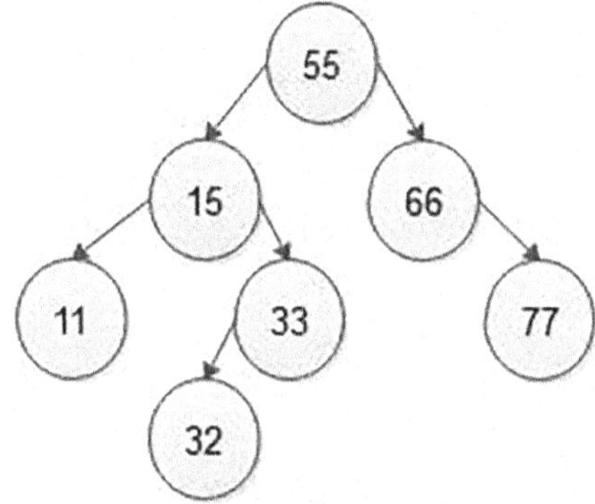

FIGURE 5.36
Before deletion.

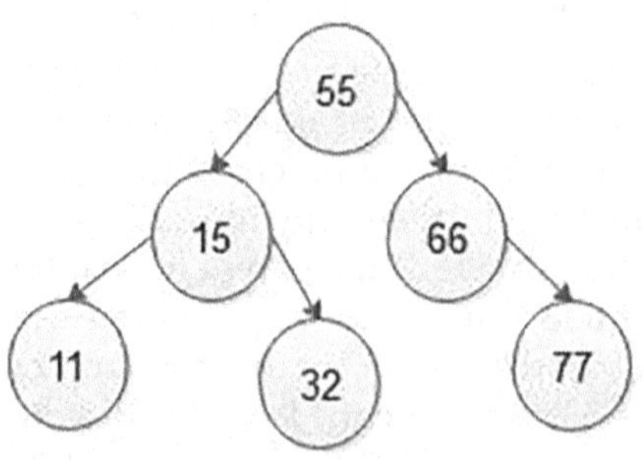

FIGURE 5.37
After deletion of node 33 from binary search tree.

5.5.2.2.3 Deletion of a Node Having Two Children

If the node which we want to delete from the BST is an interior node having two children, then find its parent node and in-order successor of the node which we want to delete. Then simply replace the in-order successor node value by deleting the node value. Then free the in-order successor node of the BST. In-order successor means to traverse the BST from a deleted node in the in-order traversing method and find the next element after deleting an element that is your in-order successor. Now take an example:

Because the node with the value 15 has two children in the above BST, the node 15 that we want to delete means will first be searched in the BST for its parent node, which is 55. Then find its in-order successor that is shown in Figure 5.38, 15 node's in-order successor is 15, **32** and 33. That is the in-order successor of 15 nodes is 32 so replace 15 node value by 32 node value and then free node 32 from the BST as is shown in Figure 5.39.

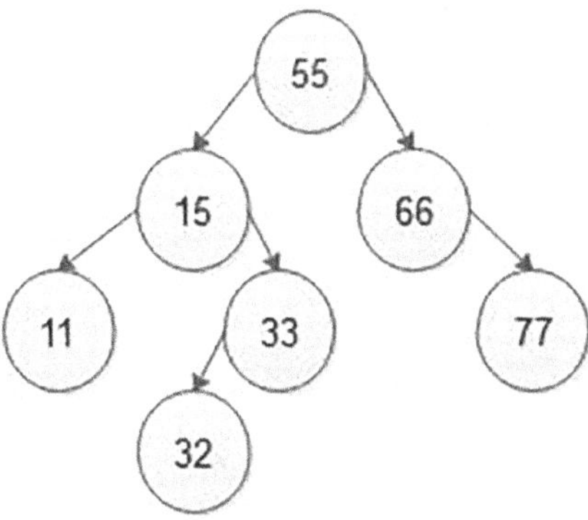

FIGURE 5.38
Before deletion of a node having two children.

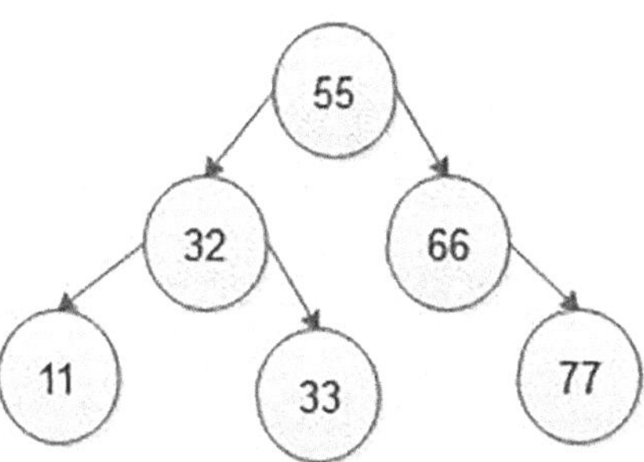

FIGURE 5.39
After deletion of node 15 from a binary search tree having two children.

5.5.2.3 Searching for a Particular Node in BSTs

When searching, the node we would like to search for is called a key node. The key node will be matched with each node starting from the root node if the value of a key node is greater than the current node, then we search for it on the right sub-tree otherwise on the left sub-tree. If we reach the leaf node and still we do not get the value of the key node, then we declare that "node is not present in the binary search tree" (Figure 5.40).

If we want to search node 33, then first compare key node 33 with root 55. Here, 33 < 55, then search the value in the left sub-tree. Now compare 33 with 15, here 33 > 15 so search key node 33 in the right sub-tree. Then compare key node 33 with the right side node of 15, that is, 33, which means 33 = 33, that is, key node 33 can be searched. Then give the message that key node 33 is found in the BST.

Program: Implementation of BST and perform insertion, deletion, searching, display operations on BST.

```
#include<stdio.h>
#include<stdlib.h>
struct bst
{
struct bst *left;
int data;
struct bst *right;
};
//Function prototypes
void insert(struct bst * root , struct bst * new1);
void in-order(struct bst *);
struct bst * search(struct bst * root, int key, struct bst **
parent);
void delet(struct bst *, int key);
//main function definition
```

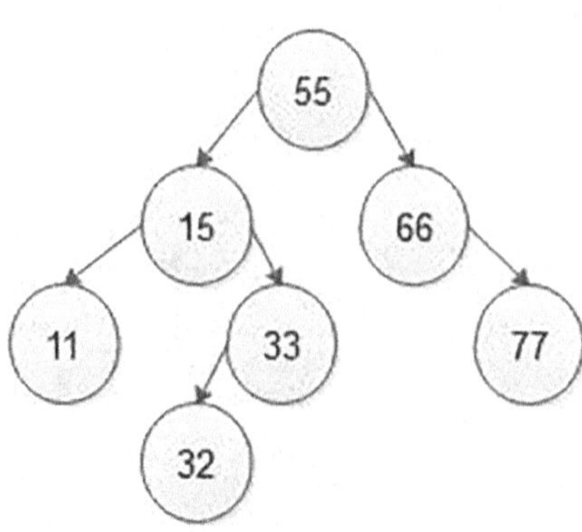

FIGURE 5.40
Binary search tree.

```
int main()
{
int choice, key;
char ch;
struct bst *root,*new1,*temp,*parent;
root=NULL;
parent=NULL;
printf("\n Binary Search Tree:");
do
{
printf("1.insert\t 2.display \t 3.search \t 4.delete \t 5.exit
\n ");
printf(" Enter your choice");
scanf("%d", &choice);
switch(choice)
{
case 1:
do
{
new1=(struct bst *)malloc(sizeof(struct bst));
printf(" Enter data");
scanf("%d",&new1->data);
new1->left=NULL;
new1->right=NULL;
if(root==NULL)
{
root=new1;
}
else
{
insert(root,new1);
}
printf("\n Do you want to insert another node[y/n]: ");
ch=getche();
}while(ch=='y'|| ch=='Y');
break;
case 2:
if(root==NULL)
printf("\n First create Binary Search Tree");
else
in-order(root);
break;
case 3:
if(root==NULL)
printf("\n First create Binary Search Tree");
else
{
printf("\n Enter data which you want to search");
scanf("%d", &key);
temp=search(root, key, &parent);
```

```
printf("\n%d", temp->data);
}
break;
case 4:
if(root==NULL)
printf("\n First create Binary Search Tree ");
else
{
printf("\n Enter data which you want to delete");
scanf("%d", &key);
delet(root, key);
}
break;
case 5:
exit(0);
default:
printf("\n Wrong choice");
}
printf("\n Do you want to continue binary search tree[y/n]");
fflush(stdin);
scanf("%c",&ch);
}while(ch=='y' || ch=='Y');
return 0;
}
//insert an element in a binary search tree
void insert(struct bst *root, struct bst *new1)
{
if(root->data<new1->data)
{
        if(root->right==NULL)
        {
        root->right=new1;
        }
        else
        {
        insert(root->right, new1);
        }
}
if(root->data>new1->data)
{
        if(root->left==NULL)
        {
        root->left=new1;
        }
        else
        {
        insert(root->left, new1);
        }
}
```

```
}
//search a node in a binary search tree and return address of
searched node and address of its //parent node
struct bst * search(struct bst *root, int key, struct bst
**parent)
{
struct bst *temp;
temp=root;
while(temp!=NULL)
{
      if(temp->data==key)
      {
      printf("\n Node is found");
      return temp;
      }
      else
      {
      *parent=temp;    //update address of parent node in
      calling function also
      if(temp->data<key)
      {
      temp=temp->right;
      }
      else
      {
      temp=temp->left;
      }
      }    // completion of outer else statement
} //completion of while loop

printf("\n Node is  not found");
return NULL;    //if node is not present in binary search tree
then return NULL address
}  // completion of search function
//delete a node from binary search tree
void delet(struct bst *root, int key)
{
struct bst * parent, *temp, *temp_succ;
temp =search(root, key, &parent);
if(temp == NULL)
{
printf("\n Node is not found in Binary search tree");
return ;
}
if(temp == root)
{
printf("\n Node which you want to delete is the root node and
we cannot delete root");
```

```
return ;
}
//deleting a node having two children
if( (temp->right!=NULL) && (temp->left!=NULL) )
{
parent=temp;
temp_succ=temp->right;
while(temp_succ->left!=NULL)   //find in-order successor of
temp node
{
parent=temp_succ;
temp_succ=temp_succ->left;
}
temp->data=temp_succ->data;  //copy data of in-order successor
node data into temp node
if(temp_succ==temp->right)    //if in-order successor node is
right side of temp node
{
parent->right=NULL;
}
else      //if in-order successor node is left side of temp
node
{
parent->left=NULL;
}
free(temp_succ);      //free in-order successor node
temp_succ = NULL;
}
//deleting a node having only right child
else if((temp->right!=NULL)&&(temp->left==NULL))
{
if(parent->right==temp)   //find whether temp node is left or
right side of parent node
{
parent->right=temp->right;
}
else
{
parent->left=temp->right;
}
temp->right=NULL;
temp->left=NULL;
free(temp);
temp=NULL;
}
```

```
//deleting a node having only left child
else if((temp->right==NULL)&&(temp->left!=NULL))
{
if(parent->left==temp)  //find whether temp node is left or
right side of parent node
{
parent->left=temp->left;
}
else
{
parent->right=temp->left;
}
temp->right=NULL;
temp->left=NULL;
free(temp);
temp=NULL;
}
//deleting a node having no child that is leaf node
else if((temp->right==NULL)&&(temp->left==NULL))
{
if(parent->left==temp)  //find whether temp node is left or
right side of parent node
{
parent->left=NULL;
}
else
{
parent->right=NULL;
}
free(temp);
temp=NULL;
}
printf("\n Deleted node in BST = %d", key);
}
//display binary search tree in in-order traversing method
void in-order(struct bst *temp)
{
if(temp!=NULL)
{
in-order(temp->left);
printf("\n%d",temp->data);
in-order(temp->right);
}
}
```

Output:

```
Binary Search Tree:1.insert     2.display       3.search        4.delete
 Enter your choice1
Enter data21

Do you want to insert another node[y/n]: y Enter data 31

Do you want to insert another node[y/n]: y Enter data 71

Do you want to insert another node[y/n]: y Enter data 51

Do you want to insert another node[y/n]: y Enter data 91

Do you want to insert another node[y/n]: y Enter data 11

Do you want to insert another node[y/n]: y Enter data 81

Do you want to insert another node[y/n]: n
Do you want to continue binary search tree[y/n]y
1.insert         2.display       3.search        4.delete        5.exit
 Enter your choice2

11
21
31
51
71
81
91
Do you want to continue binary search tree[y/n]y
1.insert         2.display       3.search        4.delete        5.exit
 Enter your choice3

Enter data which you want to search51
```

5.6 BST Algorithms

1. Create BST algorithm:
2. Insert new node in a BST algorithm:
3. Search node in a BST algorithm:
4. Delete any node from a BST algorithm:
5. In-order traversing in a BST algorithm:

5.6.1 Create BST Algorithm

Step 1: Start

Step 2: First, create a structure for the BST node as follows:

```
struct bst
{
struct bst *left;      //Store address of left sub-tree
int data;                    // Store the data in a node
struct bst *right;  // Store address of right sub-tree
};
```

Step 3: Initialize the structure's pointer type of variables such as root, parent, new1 and temp as a NULL.

Step 4: Create a new node using dynamic memory management as

```
new1=( struct bst * )malloc (sizeof (struct bst) );
```

Here, malloc() function reserves the block of memory required for the new node and that address is stored into the new1 pointer type of variable.

Step 5: Then insert new data into the data part of the newly created node also make right and left pointer as NULL using,

```
printf("Enter data into new node");
scanf("%d", &new1->data);
new1->left=NULL;
new1->right=NULL;
```

Step 6: Check whether or not the root node has a NULL address. If the root contains a NULL address, then assign the address of the new node to the root.

Step 7: Otherwise, if the root node contains some address other than NULL, then, call to insert function and pass root node's address and newly created node's address as an argument.

Step 8: Ask the user to insert more nodes in the BST.

Step 9: If the user said yes, then repeat steps 4–8.

Step 10: If the user said no, then stop creating BST.

Step 11: Stop

5.6.2 Insert New Node in BST Algorithm

Algorithm insert(struct bst *root, struct bst *new1)

Step 1: Start

Step 2: Check the data in the root node and in the new node. If the root node's data are less than the new node's data, then go to the right sub-tree. In addition, check whether or not the root node's right pointer contains the NULL address. If the root node's right pointer contains the NULL address, then attach a new node to the current root node's right. Otherwise, call the insert function recursively with the new root node's address and address of the newly created node. Here, the current root node contains the address of the right sub-tree. Here, the address of newly created node's address remains the same, which is shown as follows:

```
if(root->data < new1->data)
{
        if(root->right==NULL)
        {
        root->right=new1;
        }
        else
        {
        insert(root->right, new1);
        }
}
```

Step 3: Check the data in the root node and in the new node. If the root node's data is greater than the new node's data, then go to the left sub-tree. In addition, check whether the root node's left pointer contains the NULL address or not. If the root node's left pointer contains the NULL address, then attach a new node to the current root node's left. Otherwise, call the insert function recursively with the new root node's address and address of the newly created node. Here current root node contains the address of the left sub-tree. Here, the address of newly created node's address remains the same, which is shown as follows:

```
if(root->data > new1->data)
{
        if(root->left==NULL)
        {
        root->left=new1;
        }
        else
        {
        insert(root->left, new1);
        }
}
```

Step 4: If the new node is placed in its correct position, then stop calling the insert function recursively.

Step 5: Stop

5.6.3 Search Node in BST Algorithm

Algorithm struct bst * search (struct bst *root, int key, struct bst **parent)

Step 1: Start

Step 2: Check whether a BST is created or not.

```
If root=NULL, then
```

Print " First create Binary Search Tree."

Otherwise

Take the key from the user whom you want to search in a BST.

Step 3: Copy address of root node in temp pointer for traversing purpose.

```
temp = root
```

Step 4: Check the temp pointer's data part with the key, which you want to search till the temp pointer contains the NULL address. If temp pointer's data and key are matched, then display the message "Node is found" and return the address of the searched node.

Step 5: If temp pointer's data and key are not matched, then copy the address of current temp into parent pointer as follows:

```
*parent=temp;
```

Step 6: If temp pointer's data are less than the key value, then assign the address of the right sub-tree to the temp pointer.

```
if (temp->data<key)
{
temp=temp->right;
}
```

Step 7: Otherwise assign the address of the left sub-tree to the temp pointer.

```
temp=temp->left;
```

Repeat steps 4–7 till the temp pointer contains a NULL address.

Step 8: If the temp pointer contains the NULL address, then print "Node is not found."

Step 9: Stop.

5.6.4 Delete Any Node from BST Algorithm

Algorithm delet (struct bst *root, int key)

Step 1: Start

Step 2: Check whether or not a BST is created.

```
If root=NULL, then
```

Print "First create Binary Search Tree."

Otherwise

Take the key from the user whom you want to delete from the BST.

Step 3: Call to search function. After the execution of the search function returns two addresses, one address of node that we want to delete in temp pointer and another address of parent node which we want to delete into parent pointer.

Step 4: First check whether or not the temp pointer is NULL. If the temp pointer is NULL, then print the message "Node is not found in Binary search tree" and return to the calling function.

Step 5: If the temp pointer is not NULL, then check whether a temp pointer contains the address of the root node. If temp contains the address of root node, then print message "Node which you want to delete is a root node and we cannot delete root" and return to the calling function.

Step 6: If temp pointer does not contain addresses of the root node, then check node that we want to delete having two children, if this condition is true, then found its in-order successor and replace the deleted node value by in-order successor node's value and free the in-order successor node from the BST.

```
if( (temp->right!=NULL) && (temp->left!=NULL) )
{
parent=temp;
temp_succ=temp->right;
while(temp_succ->left!=NULL)    //find in-order successor
of temp node
{
parent=temp_succ;
temp_succ=temp_succ->left;
}
temp->data=temp_succ->data;  //copy data of in-order
successor node data into temp
     //node
if(temp_succ==temp->right)     //if in-order successor
node is right side of temp node
{
parent->right=NULL;
}
else       //if in-order successor node is left side of
temp node
{
parent->left=NULL;
}
free(temp_succ);      //free in-order successor node
temp_succ = NULL;
}
```

Step 7: If the node that we want to delete does not contain two children, then check whether the node contains the right child only. If this condition is true, then find whether the temp node is the left or right side of the parent node. If the temp node is the right side of the parent node, then assign the address of the temp node's right pointer address to the parent node's right pointer address. Otherwise, assign the address of the temp node's right pointer address to the parent node's left pointer address. In addition, free the temp node.

```
else if((temp->right!=NULL)&&(temp->left==NULL))
{
if(parent->right==temp)    //find whether temp node is
left or right side of parent node
{
parent->right=temp->right;
}
```

```
else
{
parent->left=temp->right;
}
temp->right=NULL;
temp->left=NULL;
free(temp);
temp=NULL;
}
```

Step 8: If the node that we want to delete does not contain only the right child, then check whether the node contains the left child only. If this condition is true, then find whether the temp node is the left or right side of the parent node. If the temp node is the right side of the parent node, then assign the address of the temp node's left pointer address to the parent node's right pointer address. Otherwise, assign the address of the temp node's left pointer address to the parent node's left pointer address. In addition, free the temp node.

```
else if((temp->right==NULL)&&(temp->left!=NULL))
{
if(parent->left==temp)  //find whether temp node is left
or right side of parent node
{
parent->left=temp->left;
}
else
{
parent->right=temp->left;
}
temp->right=NULL;
temp->left=NULL;
free(temp);
temp=NULL;
}
```

Step 9: If the node that we want to delete does not contain only the left child, then check whether the node does not contain both left and right child. If this condition is true, then find whether the temp node is the left or right side of the parent node. If the temp node is the left side of the parent node, then the parent node's left pointer is assigned address as NULL, otherwise the parent node's right pointer is assigned address as NULL. In addition, free the temp node.

```
else if((temp->right==NULL)&&(temp->left==NULL))
{
if(parent->left==temp)  //find whether temp node is left
or right side of parent node
{
parent->left=NULL;
}
else
{
```

```
parent->right=NULL;
}
free(temp);
temp=NULL;
}
```

Step 10: After checking all conditions, the print value of the deleted node.

Step 11: Stop

```
//display binary search tree in in-order traversing
method
void in-order(struct bst *temp)
{
if(temp!=NULL)
{
in-order(temp->left);
printf("\n%d",temp->data);
in-order(temp->right);
}
}
```

5.6.5 In-order Traversing of BST Algorithm

In-order traversing of the BST T with root R.

Algorithm in-order (root R)

1. Traverse the left sub-tree of root R, that is, call in-order (left sub-tree)
2. Visit the root, R and process.
3. Traverse the right sub-tree of root R, that is, call in-order (right sub-tree)

5.7 Applications of Binary Tree and BST

1. BST is used to implement dictionary.
2. BST is used to implement multilevel indexing in the database.
3. BST is used to implement the searching algorithm.
4. Heaps is a special type of binary tree: Used in implementing efficient priority queues, which in turn are used for scheduling processes in many operating systems, quality-of-service in routers and the path-finding algorithm used in artificial intelligence applications, including robotics and video games. It is also used in heap-sort.
5. Syntax tree: In the syntax analysis phase of the compiler, parse tree or syntax tree is drawn for parsing expressions either top-down

parsing or bottom-up parsing.

6. BST is used in Unix kernels for managing a set of virtual memory areas (VMAs). Each VMA represents a section of memory in a Unix process.
7. Each IP packet sent by an Internet host is stamped with a 16-bit id that must be unique for that source-destination pair. The Linux kernel uses an AVL tree or height-balanced tree for an indexed IP address.
8. BST is used to efficiently store data in a sorted form in order to access and search stored elements quickly.
9. BST is used in dynamic sorting.
10. **Huffman Code Construction:** A method for the construction of minimum redundancy codes. Huffman code is a type of optimal prefix code that is commonly used for lossless data compression. The algorithm has been developed by David A. Huffman. The technique works by creating a binary tree of nodes. The nodes count depends on the number of symbols.

The main idea is to transform plain input into variable-length code. As in other entropy encoding methods, most common symbols are generally represented using fewer bits than less common symbols.

Huffman Encoding:

A type of variable-length encoding that is based on the actual character frequencies in a given document.

Huffman encoding uses a binary tree:

1. To determine the encoding of each character.
2. To decode an encoded file that is to decompress a compressed file, putting it back into ASCII. Example of a Huffman tree for a text with only six characters:

Leaf nodes are characters.

Left branches are labeled with a 0, and right branches are labeled with a 1.

If you follow a path from root to leaf, you get the encoding of the character in the leaf.

Example: 101 = 'j' and 01 = 'm'.

Characters that appear more frequently are nearer to the root node, and thus their encodings are shorter and require less memory.

Building Huffman tree:

1. First reading through the text, determine the frequencies of each character.
2. Create a list of nodes with character and frequency pairs for each character within the text sorted by frequency.

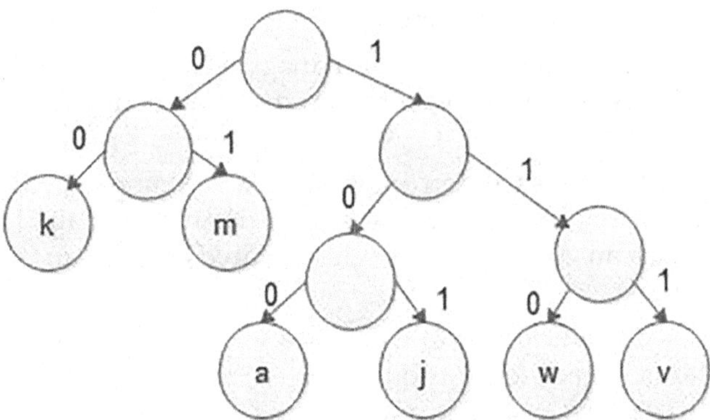

FIGURE 5.41
Huffman tree.

3. Remove and combining the nodes with the two minimum frequencies, creating a new node that is their parent.
   - The left child with the lowest frequency node.
   - The right child with the second lowest frequency node.
   - The frequency of the parent node has the sum of the frequencies of its children, which is the left and right child.
4. Add the parent to the list of nodes maintaining the sorted order of the list.
5. Repeat steps 3 and 4 as long as there is only a single node in the list, which will be the root of the Huffman tree indicated in Figure 5.41.

11. Binary trees are a good means of expressing arithmetical expressions. The leaves or exterior nodes are operands, and the interior nodes are operators. Two usual types of expressions that a binary expression tree can describe are algebraic and Boolean. These binary trees can represent expressions that incorporate both unary and binary operators.

Each node of a binary tree, that is, in a binary expression tree, has zero, one, or two children. This constricted arrangement simplifies the processing of expression trees.

5.8 Heaps

5.8.1 Basic of Heap data Structure

The heap data structure is a special type of binary tree.

Binary heap is of two types:

1. Max heap
2. Min heap

5.8.2 Definition of Max Heap

Max heap is defined as follows:

A heap of size 'n' is a binary tree of 'n' nodes that satisfies the following two constraints:

1. The binary tree is an **almost complete binary tree**. That is, each leaf node in the tree is either at level d or at level d-1, which means that only one difference between levels of leaf nodes is permitted. At depth d, that is the last level, if a node is present, then all the nodes to the left of that node should also be present first and no gap between the two nodes from left to the right side of level d.
2. Keys in nodes are arranged such that the content or value of each node is less than or equal to the content of its father or parent node, which means for each node info[i] <= info[j], where j is the father of node i. This heap is called as max heap.

Here, in heap, each level of binary tree is filled from left to right, and a new node is not placed on a new level until the preceding level is full.

In a min binary heap, the key at the root must be minimum among all keys present in Binary Heap. The same characteristic must be iteratively true for all nodes in a binary tree.

Examples of Max Heap (Figure 5.42).

5.8.3 Operations Performed on Max Heap

The following operations are performed on a max heap data structure:

1. Insertion or creation of heap

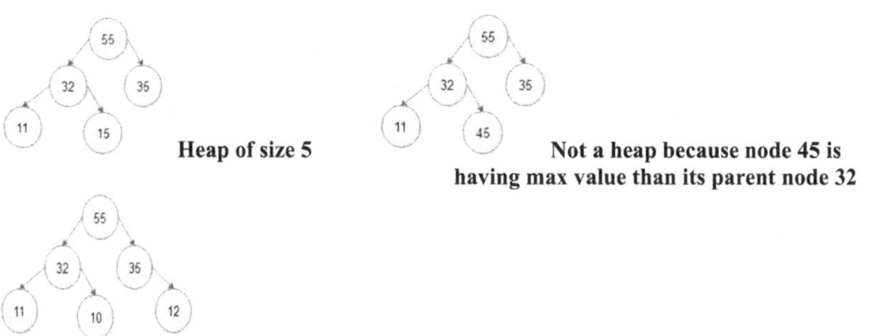

FIGURE 5.42
Examples of max heap and not max heap.

2. Deletion
3. Finding a maximum value.

5.8.3.1 Insertion Operation in Max Heap

Insertion operation in max heap is carried out using the following algorithm:

Step 1: Start

Step 2: Add the new node as the last leaf from left to right.

Step 3: Check the new node value against its parent node.

Step 4: If the new node value is greater than its parent, then swap both of them.

Step 5: Repeat steps 3 and 4 as long as the new node value is lesser than its parent node or the new node reaches the root.

Step 6: Stop

Inserting a new key takes **O (log n)** time.

Example (Figure 5.43):

The above binary tree is an almost complete binary tree and the content or value of each node is lesser than the content of its father or parent node and each level of binary tree is filled from left to right, that is, binary tree satisfying both ordering property and structural property according to the max heap data structure so above binary tree is max heap of size 5.

Now insert a new node with a value 65 in above max heap of size 5.

Step 1: Insert the **new node** with value 65 as a **last leaf** from left to right, which means the new node is added as a left child of the node with value 35. After adding 65, max heap is as follows:

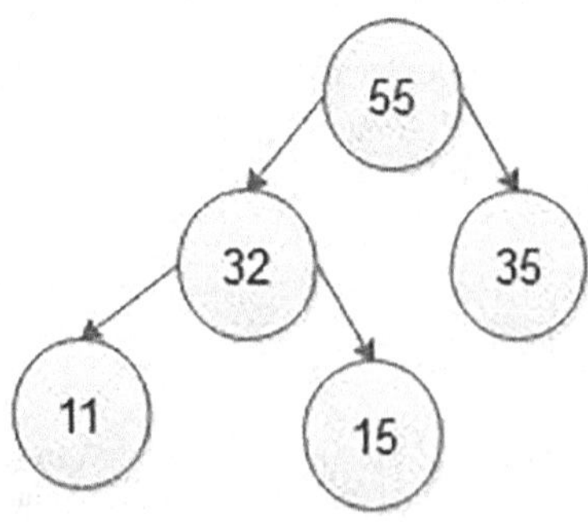

FIGURE 5.43
Max heap of size 5.

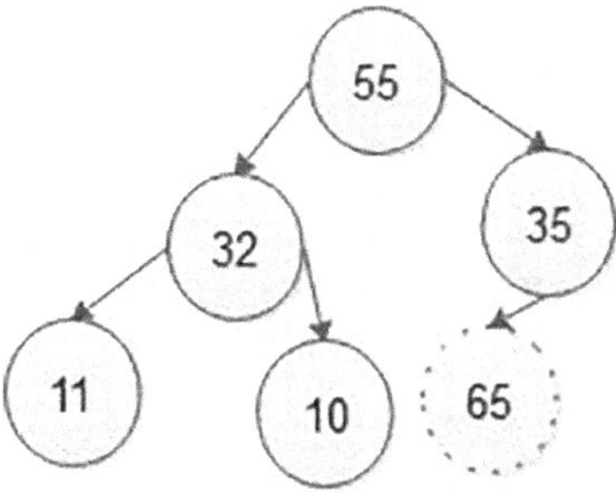

Step 2: Compare **the** new node value (65) with its parent node value (35). That means 65 > 35

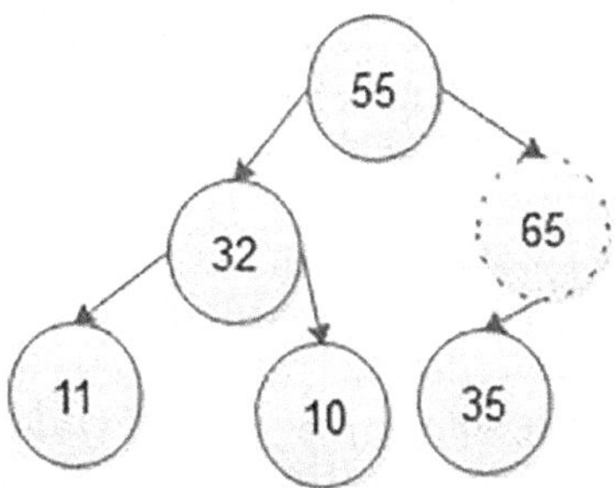

Here the new node value (65) is greater than its parent value (35), then swap both of them. After sapping, max heap is as shown in below figure:

Step 3: Now, again compare a new node value (65) with its parent node value (55). Here the new node value (65) is greater than its parent value (55), then swap both of them. After sapping, max heap is as shown in below figure:

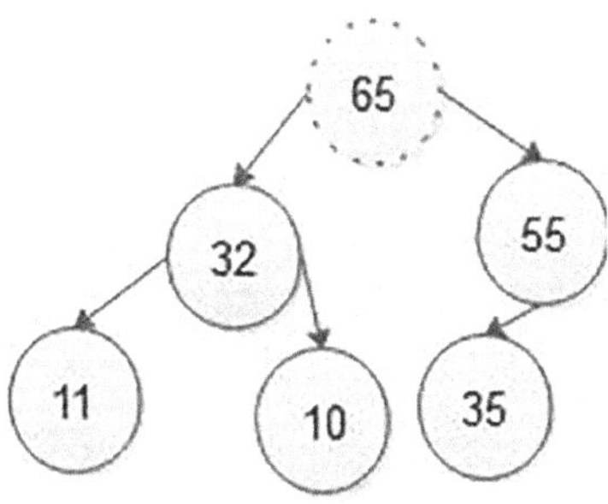

Step 4: Now new node is a root node hence stopping comparing and the final heap of size 6 is as shown in the Figure 5.44.

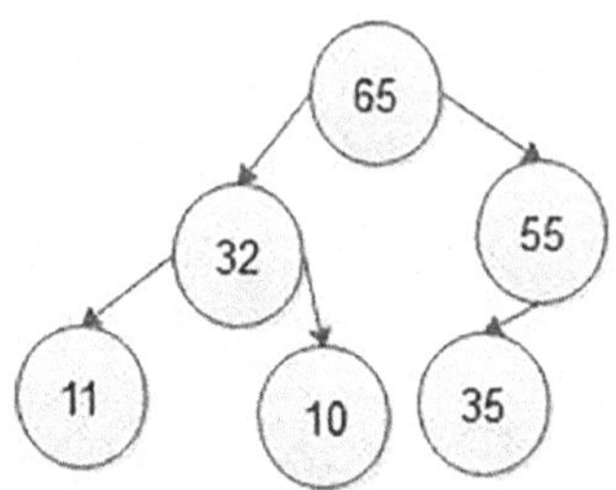

FIGURE 5.44
Max heap of size 6 after insertion of node 65.

5.8.3.2 Deletion Operation in Max Heap

In a max heap, removing the last node is very straightforward as it does not disturb max heap characteristics. Removing a root node from a max heap is challenging as it disturbs the max heap characteristics. We use the below steps to remove the root node from the max heap.

Step 1: Start

Step 2: Interchange the root node with the last node in max heap.

Step 3: After interchanging, remove the last node.

Step 4: Compare the root node value with its left child node value.

Step 5: If the root value is smaller than its left child, then compare the left child node value to its right sibling. If not, continue with step 7.

Step 6: If the left child value is larger than its right sibling, then interchange root with the left child. Otherwise, swap root with its right child.

Step 7: If the root value is larger than its left child, then compare the root value with its right child value.

Step 8: If the root value is smaller than its right child, then swap root with the right child. Otherwise, stop the process.

Step 9: Repeat the same one until the root node is set to the correct position.

Step 10: Stop.

Inserting a key takes O (log n) time.

Example (Figure 5.45):

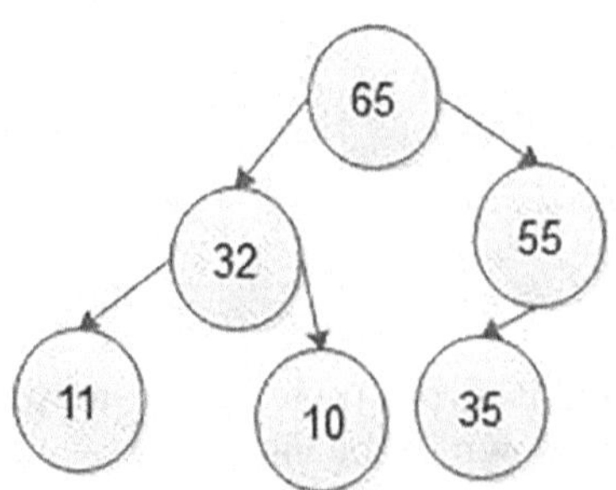

FIGURE 5.45
Max heap of size 6.

In the above max heap, if we want to delete 65 root nodes, then follow the following steps:

Step 1: Swap the root node 65 with the last node 35 in max heap. After swapping, the max heap is as follows:

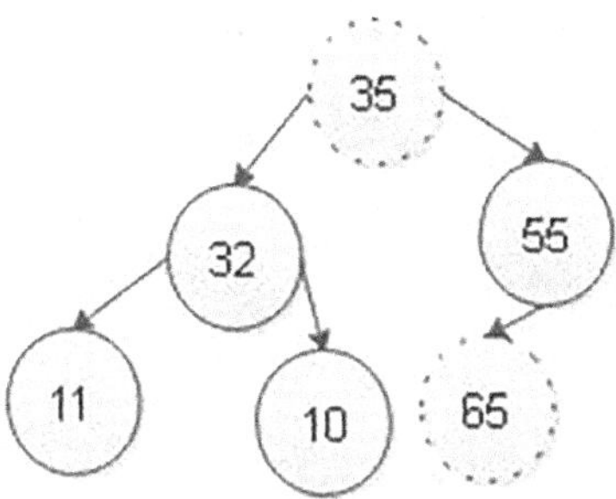

Step 2: Delete the last node that is 65 from max heap. After deleting node with value 65 from max heap, the max heap is as follows:

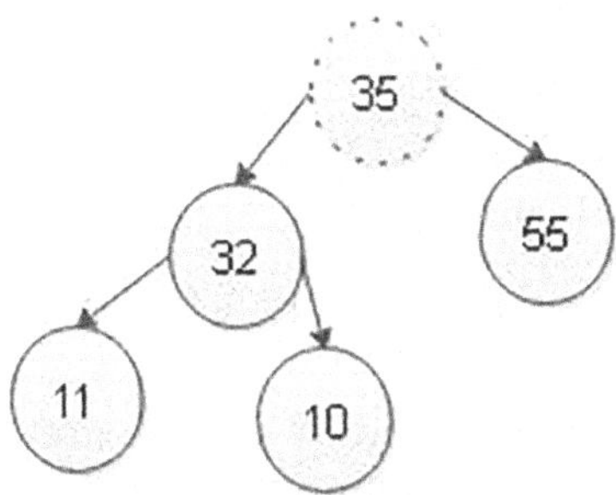

Step 3: Compare root node 35 with its left child 32. Here, the node with value 35 is larger than its left child. Therefore, we compare node with value 35 is compared with its right child 55. Here, the node with value 35 is smaller than its right child 55. Therefore, we swap both of them. After swapping, max heap is as follows:

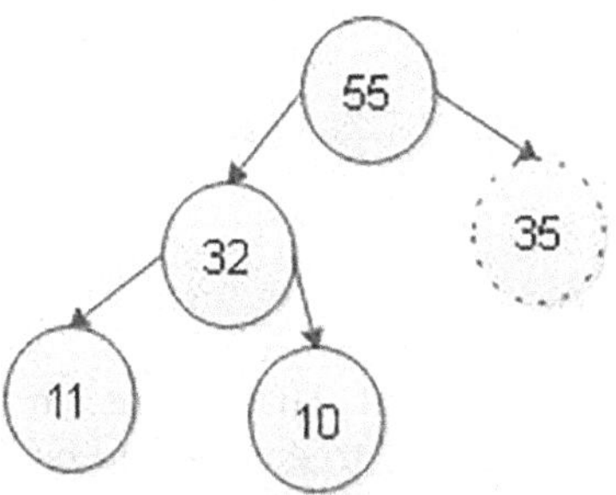

Here, the node with value 35 does not have left and right child. Therefore, we stop the process.

Finally, max heap after deleting root node 65 is as follows:

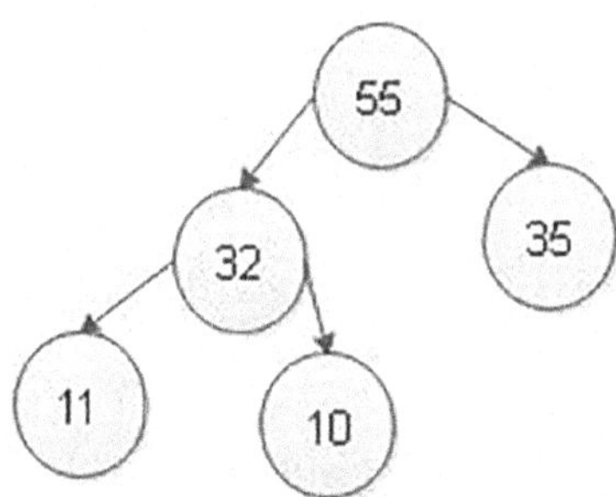

5.8.3.3 Finding Maximum Value in Max Heap

Finding the node with the maximum value in a max heap is quite simple. In max heap, the root node has the maximum value as all other nodes in the max heap. Thus, directly we can display the value of the root node as the maximum value in the max heap.

5.8.4 Applications of Heaps

1. **Heap Sort:** Heap sort uses a binary heap for sorting an array in O (n log n) time.
2. **Graph Algorithms:** The priority queues are particularly used in graph algorithms like Dijkstra's shortest path and Prim's minimum spanning tree.
3. **Priority Queue:** Priority queues may be implemented effectively using binary heap.
4. Many problems can be resolved effectively by utilizing heaps such as:
   i. Nth largest element in an array.
   ii. Sort an almost sorted array.
   iii. Merge N number of sorted arrays.

5.9 AVL Tree

5.9.1 Introduction of AVL Tree

The AVL tree is a self-balanced BST. That means, an AVL tree is also a BST, but it is a balanced tree. A binary tree is said to be balanced, if the difference between the heights of the left sub-trees and right sub-trees of every node in the tree is either –1, 0 or +1. In other words, a binary tree is said to be balanced if for every node's height of its children differs by at most one. In an AVL tree, every node maintains extra information known as **balance factor**. The AVL tree was initiated in the year 1962 by G.M. Ade'son-Ve'skii and E.M. Landis in their honor's height balanced BST is termed as AVL tree. In the complete balanced tree, left and right sub-trees of any node would have the same height. Every AVL Tree is a BST, but all the BSTs need not to be AVL trees.

5.9.2 Definition

An empty tree is height-balanced, if T is a non-empty BST with T_L and T_R as its left and right sub-tree, then T is height-balanced if,

1. T_L and T_R are height-balanced and
2. $| h_L - h_R | <= 1$.

where h_L and h_R are the heights of T_L and T_R, respectively.

An AVL tree is a balanced BST. In an AVL tree, the balance factor of every node is either –1, 0 or +1.

The balance factor of a node is the difference between the heights of the left and right sub-trees of that node. The height of the left sub-tree – height of the right sub-tree or height of the right sub-tree – height of the left sub-tree is used to compute the balancing factor of a node. In the following explanation, we calculate as follows:

Balance factor (T) = height of left sub-tree – height of right sub-tree

where T is a binary tree.

BF (T) = –1, 0, 1.

If AVL tree's balance factor is exceeding the value 1 for any of the nodes, then rebalancing has to be taken place. After each insertion and deletion of items, if the balance factor of any node exceeds value 1, then rebalancing of the tree is carried out by using different rotations. These rotations are characterized by the nearest ancestor of the newly inserted node whose balance factor becomes +2 or –2.

5.9.3 AVL Tree Rotations

In the AVL tree, after each operation such as inserting and deleting, we have to check the balance factor of each node in the tree. If each node fulfills the balance factor condition, then we conclude the operation; otherwise, we must make it balanced. We use rotation operations to balance the tree every time the tree becomes unbalanced because of any operation.

Rotation operations are used to balance a tree.

Rotation is the movement of the nodes to the left or right to balance the tree.

The following are four types of rotations:

1. **L L rotation**
2. **R R rotation**
3. **L R rotation**
4. **R L rotation**

Example:

1. **L L rotation:**

   It is used when a new node is inserted in the left sub-tree of the left sub-tree of node A.

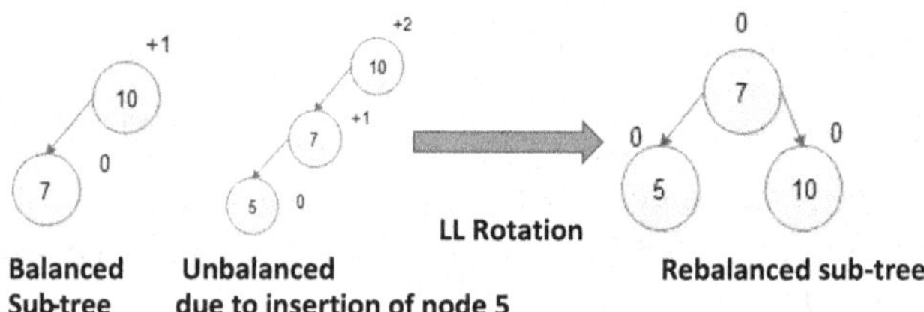

2. **R R rotation:**

   It is used when a new node is inserted in the right sub-tree of the right sub-tree of node A.

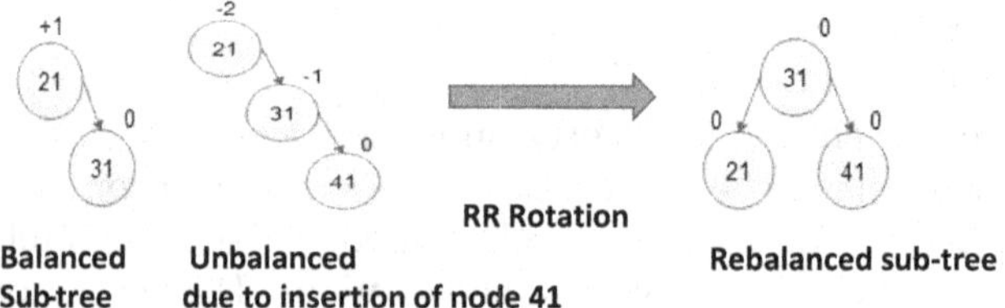

3. **L R rotation:**

   It is used when a new node is inserted in the right sub-tree of the left sub-tree of node A.

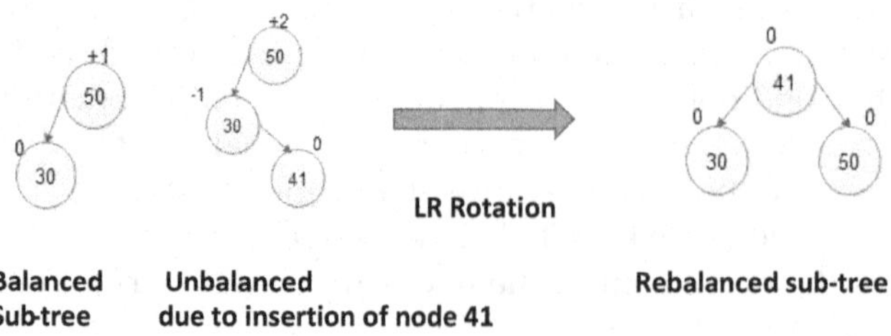

4. R L rotation:

It is used when a new node is inserted in the left sub-tree of the right sub-tree of node A.

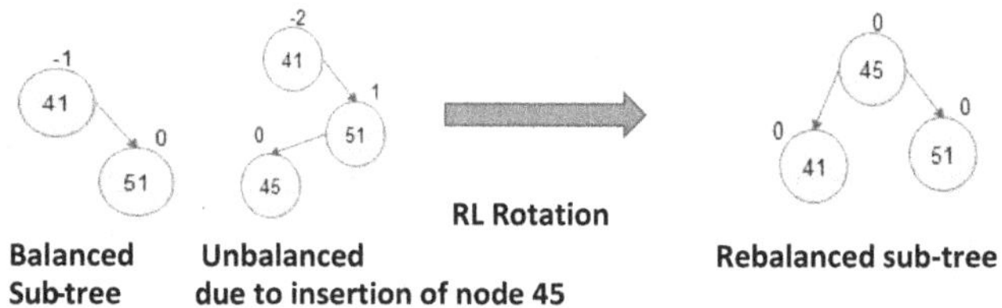

5.9.4 Operations on an AVL Tree

The following operations are carried out in an AVL tree:

1. Search a node in an AVL tree
2. Insertion of a node in an AVL tree
3. Deletion of a node from an AVL tree.

5.9.4.1 Search Operation in AVL Tree

In the AVL tree, the search operation is executed with a time complexity O (log n). The search operation is conducted in the same way as the search operation of the BST. We use the subsequent steps to search an element in an AVL tree:

Step 1: Start

Step 2: Read the search element from the user.

Step 3: Next, compare the search element to the root node value in the tree.

Step 4: If both are equal, then display "Element is Searched" and wind up the search function.

Step 5: If both are not equal, then verify whether a search element is smaller or greater than that of the parent node value.

Step 6: If the search element is smaller, then continue the search process in the left sub-tree.

Step 7: If the search element is larger, then continue the search process in the right sub-tree.

Step 8: Repeat the same as long as we find the exact element or we finish with a leaf node.

Step 9: If we reach the node with a search value, then display "Element is Searched" and terminate the function.

Step 10: If we reach a leaf node and it is also not matching, then display "Element is not Searched" and terminate the function.

Step 11: Stop.

5.9.4.2 Insertion Operation in AVL Tree

In the AVL tree, the insertion activity is carried out with O (log n) time complexity. In the AVL Tree, the new node is always inserted as a leaf node. The insertion operation is carried out as follows:

Step 1: Start

Step 2: Insert the new item in the tree with the BST insertion logic.

Step 3: Once inserted, check the balance factor for each node.

Step 4: If the balance factor of every node is 0 or 1 or –1, then go to step 6.

Step 5: If the balance factor of a node is not 0 or 1 or –1, then the tree is said to be unbalanced. Then perform the appropriate rotation technique to achieve balance. In addition, get on with the next operation.

Step 6: Stop.

Example:

Insertion of the following nodes in an AVL tree:
55, 66, 77, 15, 11 and 33

New Identifier	After Insertion	Rotation Technique	After Rebalancing
55	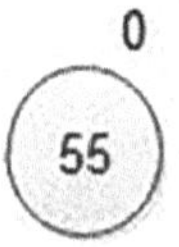 0 55		No rebalancing needed
66	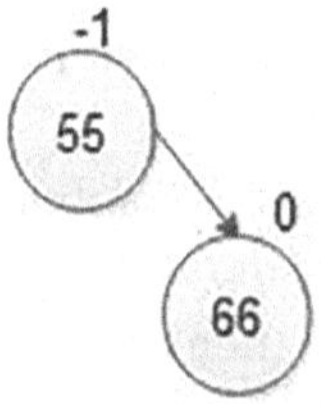 -1 55 → 0 66		No rebalancing needed
77	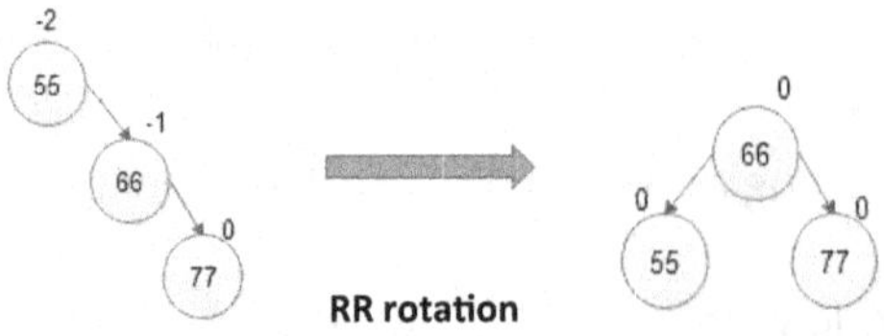 -2 55 → -1 66 → 0 77 ⇒ RR rotation ⇒ 0 66; 0 55, 0 77		

(Continued)

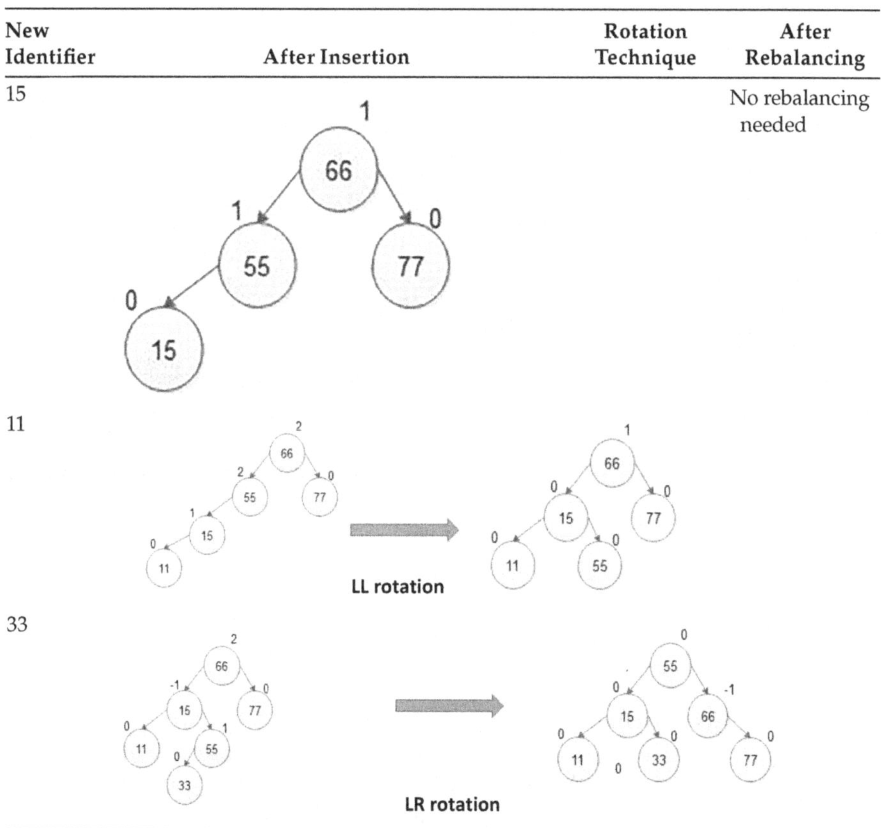

New Identifier	After Insertion	Rotation Technique	After Rebalancing
15			No rebalancing needed
11		LL rotation	
33		LR rotation	

5.9.4.3 Deletion Operation in AVL Tree

In the AVL Tree, the deletion operation is similar to the deletion operation in BST. However, after each deletion operation, we must necessary to check with the balance factor condition. If the tree is balanced after deletion, then go to the next operation otherwise perform the proper rotation to make the tree balanced.

5.9.5 Advantages

1. When compared to the number of insertion and deletions, the number of times a search operation is carried out is relatively large. The use of an AVL tree always results in superior performance.
2. The main objective here is to keep the tree balanced at all times. Such balancing causes the depth of the tree to remain at its minimum, reducing the overall cost of search.
3. AVL tree gives the best performance for dynamic tables.

5.9.6 Disadvantages

1. Insertion and deletion of any node from an AVL tree become complicated.

5.10 B Trees

5.10.1 Introduction of B Tree

B tree is not a binary tree. B tree is a balanced m-way tree that is each tree node contains a maximum of (m-1) data elements in the node and has m branches or links.

For example, a balanced four-way tree node contains three data fields and four-link fields shown as follows:

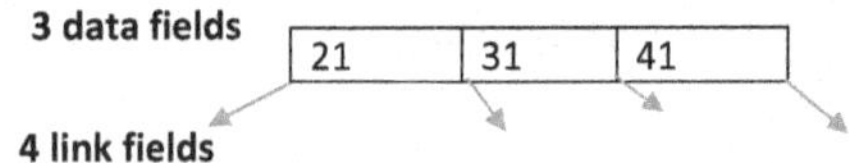

B tree finds its use in external sorting.

In a BST, AVL Tree, every node can have only one value or key and a maximum of two children, but there is some other type of search tree called B-Tree in which a node can store more than one value or key and it can have more than two children. B-Tree was developed in 1972 by Bayer and McCreight with the name Height Balanced m-way Search Tree. Later, it was termed as B-Tree.

B-Tree can be defined as follows:

The B-Tree is a self-balanced search tree with multiple keys in every node and more than two children for every node. In this case, the number of keys in a node and the number of children for a node depend on the order of the B-Tree. Every B-Tree has an order.

5.10.2 B-Tree of Order m Has the Following Properties

1. All the leaf nodes must be at the same level.
2. Each node has a maximum of m children.
3. Each node has one fever key than children with a maximum of m-1 keys or data values.
4. The keys are arranged in defining order within a node such as ascending order. In addition, all keys in the left sub-tree are less valuable than key and the right sub-tree has greater value.

5. When a new key is to be inserted into a full node, the node is split into two nodes and key with a median value is inserted into the parent node. If the parent node is root and it is likewise full, then a new root is created.

For example, B-Tree of Order 4 contains maximum three key values in a node and maximum four children for a node.

5.10.3 Operations on a B-Tree

1. **B-Tree Insertion:**
2. **B-Tree Deletion:**
3. **Search Operation in B-Tree:**

5.10.3.1 B-Tree Insertion

In a B-Tree, the new element should only be added to the leaf nodes. This means that the new value of the key remains attached to the leaf node only. The insertion operation takes place as follows:

Step 1: Start

Step 2: First verify if the tree is empty.

Step 3: If the tree is empty, create a new node with a new key value and place it in the tree as the root node.

Step 4: If the tree is not empty, then discover a leaf node to which the new key value can be inserted using the BST logic.

Step 5: If that leaf node has an empty place, then add the new key value for that leaf node by preserving ascending order of the key value as part of the node.

Step 6: If that leaf node is already full, then split that leaf node by sending a middle value to its parent node. Repeat the same until sending value is fixed into a node.

Step 7: If the splitting has taken place of the root node, then the middle value becomes a new root node in the tree, and the height of the tree is increased by one.

Step 8: Stop.

Example:

Consider a building of B-tree of order or degree 4 that is a balanced four-way tree where each node can hold three data values and have four links or branches.

Suppose we want to insert the following data values in B-tree,

1, 5, 6, 2, 8, 11, 13, 18, 20, 7, 9.

Steps:

First, value 1 is placed in a node, which can also accommodate next two values, that is, 5 and 6.

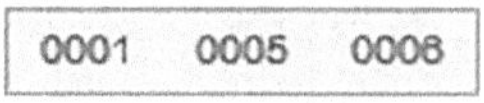

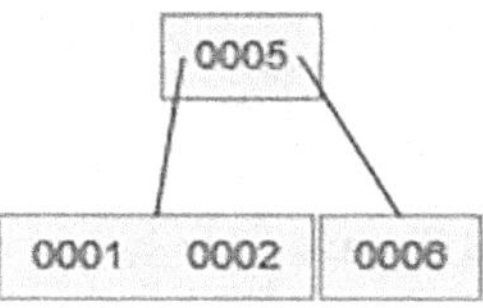

When the fourth value 2 is to be added, the node is split at a median value 5 into leaf nodes with a parent at 5.

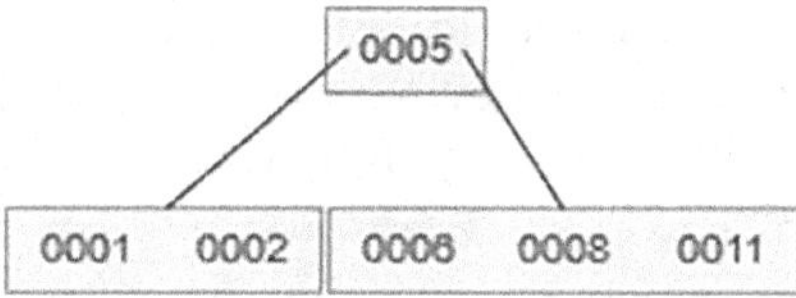

Then, 8 and 11 are added to a leaf node.

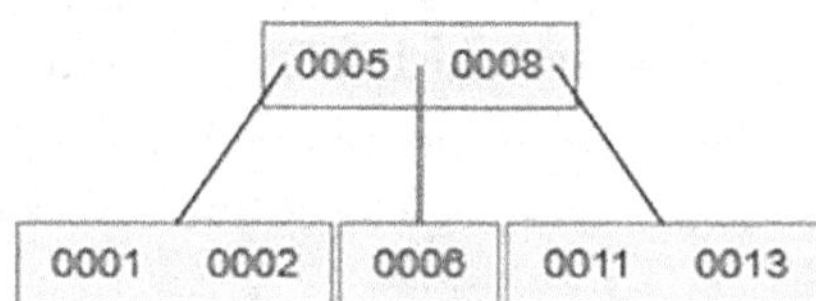

13 is to be added. However, the right leaf node is full. Therefore, it is split at a median value 8 and thus 8 moves up to the parent. In addition, the right leaf node is split into nodes.

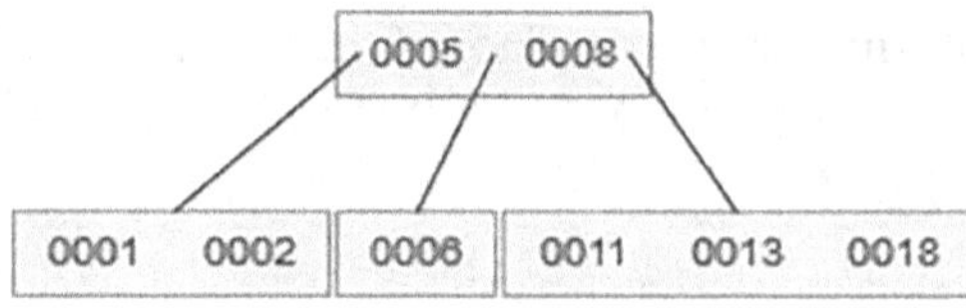

18 is to be added, which is shown as follows:

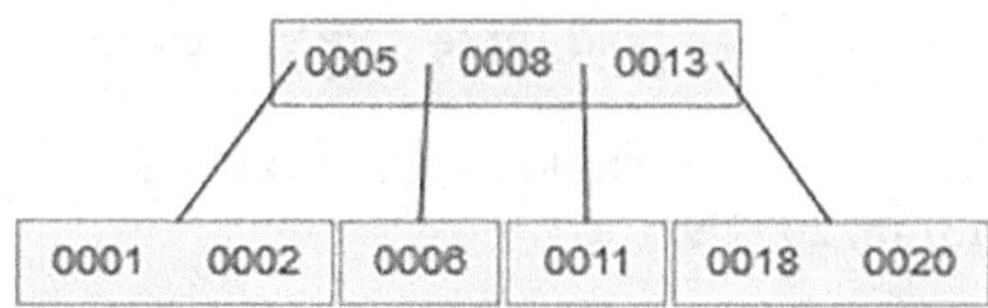

20 is to be added, which is shown as follows:

0005 0008 0013

0001 0002 | 0006 | 0011 | 0018 0020

Next 7 and then 9 is to be added, which is shown as follows:

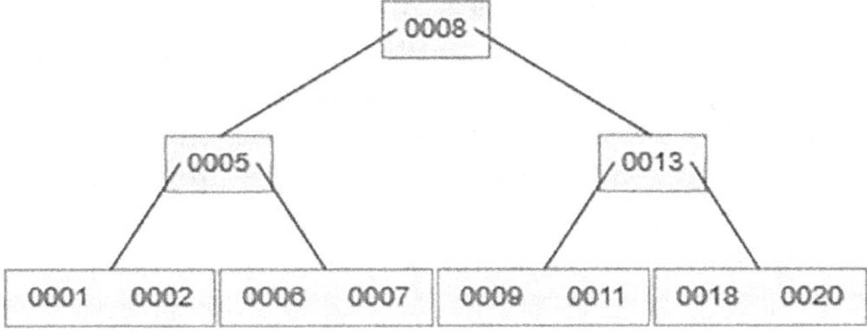

5.10.3.2 B-Tree Deletion

The delete method lets you search for the first record you want to delete.

1. If the record is in the terminal node, the deletion is straightforward. The record along with the appropriate pointer is deleted.
2. If the record is not in the terminal node, it is replaced by a copy of its successor, which is a record with a higher subsequent value.

Example:

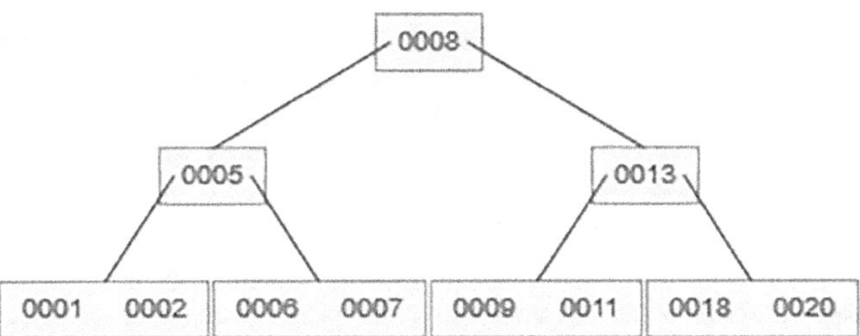

If 20 is searched, then delete 20 from that node. After the deletion, B-tree is like as follows:

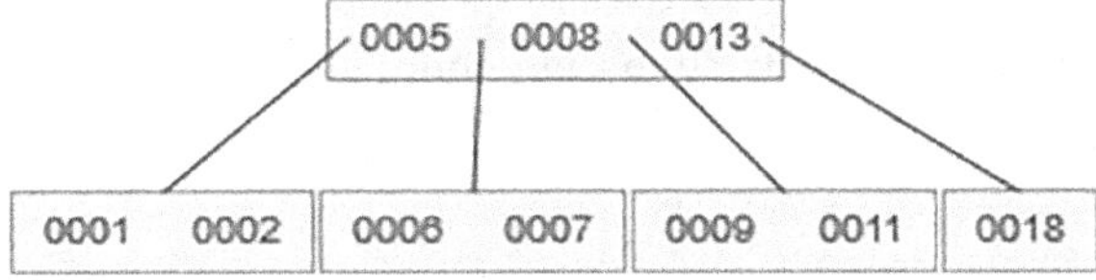

If element 5 is deleted, then find its successor node element that is 6 and replace 5 with 6.

After deletion of element 5, B-tree becomes as follows:

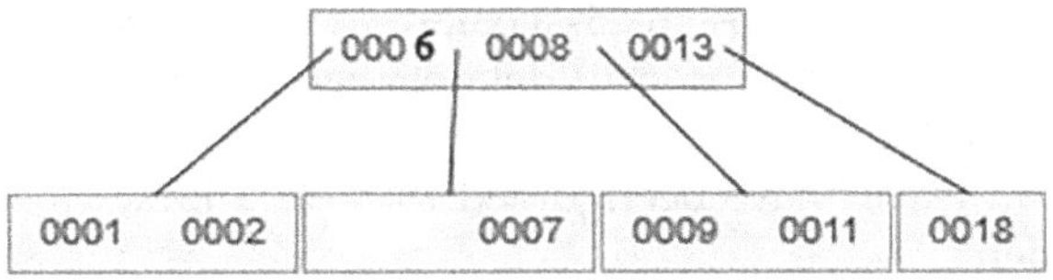

5.10.3.3 Search Operation in B-Tree

In a B-tree, the search function is the same as that of BST. In a BST, the search process begins from the root node and the rest of time we make a two-way decision either we go to the left sub-tree or we go to the right sub-tree. In a B-Tree, the search process starts from the root node, but every time we make an n-way decision where n is the total number of children that node has. In a B-tree, the search operation is performed with **O (log n)** time complexity. The search procedure is carried out as follows:

Step 1: Start.

Step 2: First, read the search element from the end user.

Step 3: Then compare, the search element with the first key value of the root node in the tree.

Step 4: If both are equal, then display "Given node is searched" and wind up the function.

Step 5: If both are not matching, then verify whether a search element is smaller or larger than that key value.

Step 6: If the search element is smaller, then continue the search process in the left sub-tree.

Step 7: If the search element is larger, then compare with the next key value in the same node and repeat steps 3, 4, 5 and 6 until we found an exact match or comparison completed with the last key value in a leaf node.

Step 8: If we completed with last key value in a leaf node, then display "Element is not found" and terminate the function.

Step 9: Stop.

5.10.4 Uses of B-tree

1. External sorting purpose B-tree is used.
2. B tree is used in a database management system.

5.11 B+ Trees

5.11.1 Introduction of B+ Tree

B+ tree is not a binary tree. B+ tree is a balanced m-way tree, that is, each tree node contains a maximum of (m-1) data elements in the node and have m branches or links.

In B+ tree, all leaves have been joined to form a linked list of keys in a sequential order.

B+ tree has two parts:

1. The index part is interior nodes.
2. The sequence set is the leaves or exterior nodes.

The linked leaves are a brilliant aspect of B+ tree. Keys can be accessed effectively both directly and sequentially. B+ tree is used to offer indexed sequential file organization. The key values in the sequence set are the key values in the record collection. The key values in the index part exist solely for internal purposes of direct access to the sequence set.

In the B+ tree, + indicates extra functionality than that of the B-tree, that is, leaf nodes are connected.

5.11.2 B+ Tree of Order m Has the Following Properties

1. All the leaf nodes must be at the same level.
2. Each node has a maximum of m children or link parts.
3. Each node has one fever key than children with a maximum of m-1 keys or data values.
4. The keys are arranged in a defining order within a node such as ascending order. In addition, all keys in the left sub-tree are less value than key and the right sub-tree has a greater value.
5. When a new key is to be inserted into a full node, the node is split into two nodes, and key with a median value is inserted into the parent node. If the parent node is root and it is likewise full, then a new root is created.
6. All leaves have been connected to form a linked list of keys in sequential order.

For example, the B+ Tree of order 3 contains maximum two key values in a node and maximum three children or link part of a node.

5.11.3 Operations on a B+ Tree

1. **B+ Tree Insertion:**
2. **B+ Tree Deletion:**
3. **Search Operation in B+ Tree:**

5.11.3.1 B+ Tree Insertion

B+ tree insertion algorithm:

In a B+ Tree, the new element must be added only at leaf nodes. That means, always the new key value is attached to the leaf node only. The insertion operation is performed as follows:

Step 1: Start.

Step 2: First verify if the tree is empty.

Step 3: If the tree is empty, create a new node with a new key value and place it in the tree as the root node.

Step 4: If the tree is not empty, then discover a leaf node to which the new key value can be added using BST logic.

Step 5: If that leaf node has an empty position, then add the new key value for that leaf node by preserving ascending order of key value within the node.

Step 6: If that leaf node is already full, then split that leaf node by sending the median value to its parent node. Repeat the same until sending value is fixed into a node.

Step 7: If the splitting is occurring to the root node, then the median value becomes a new root node in the tree and the height of the tree is increased by one.

Step 8: After inserting a new key value into B+ tree, all leaves have been connected to form a linked list of keys in sequential order.

Step 9: Stop.

Example:

Consider a building of the B+ tree of order or degree 3 that is a balanced three-way tree where each node can hold two data values and have three links or branches.

Suppose we want to insert the following data values in the B+-tree,

10, 20, 30, 12, 21, 55 and 45.

Here, the B+ tree of order 3 having two data part. Therefore, we first insert 10 in a node then 20 data at the next right empty field in node, then B+ tree looks like as follows:

0010	0020

After the insertion of values 30 in the B+ tree node, split, then node having median 20 and make that node as a parent node and insert 20 and 30 value into one node and 10 values in another node and now these two nodes become leaf nodes, so connect that two nodes using a linked list. After inserting value 30, B+ tree looks like as follows:

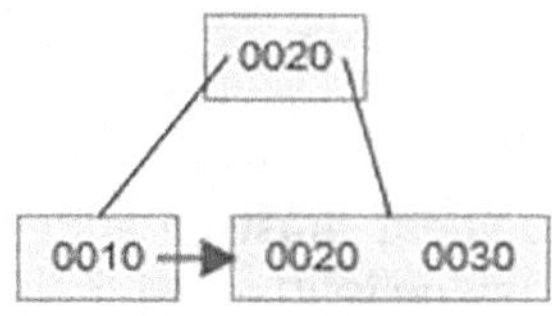

After insertion of 12 value in B+ tree,

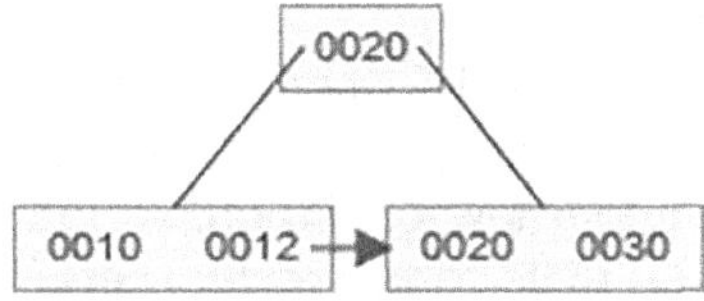

After insertion of 21 value in B+ tree,

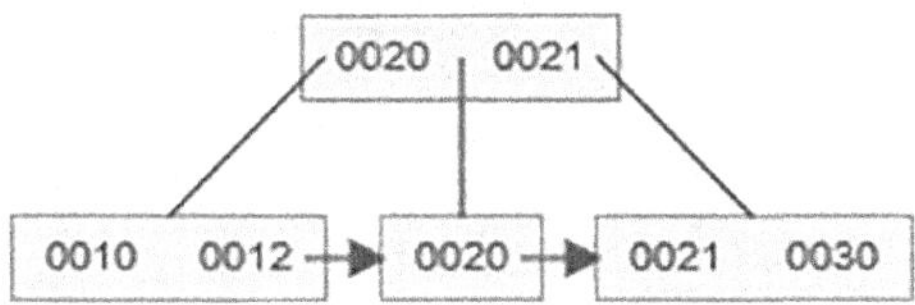

After insertion of 55 value in B+ tree,

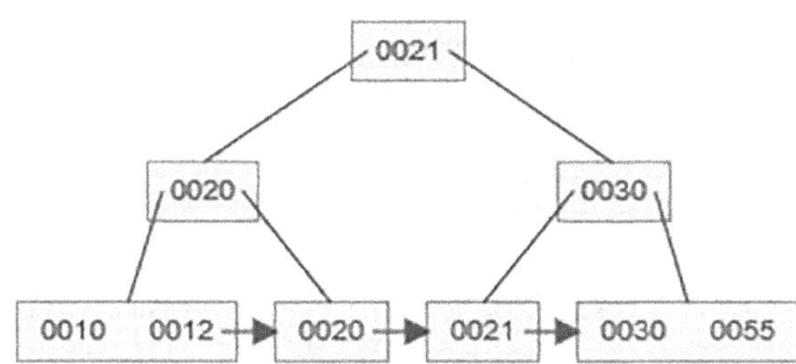

After insertion of 45 value in B+ tree,

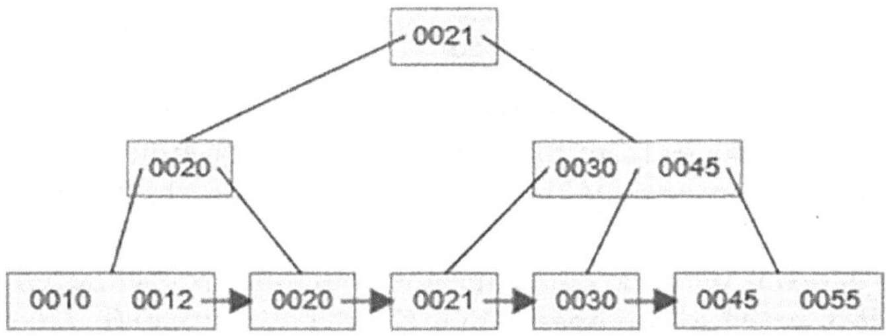

5.11.3.2 B+ Tree Deletion

B+ tree deletion algorithm:

1. If the deleted key is the indexed node's value, then delete the index node's key and data from the leaf node.
2. If the leaf node underflows, merge with siblings and delete key in between them.

3. If the index node underflows, merge with siblings and move down key in between them.

Deletion of 30 values from above B+ tree, then 30 values are in both leaf node and index node. Therefore, delete the value from both leaf and index node. Then index node goes underflow condition, so the successor key value 55 becomes median and goes to its parent node. Therefore, 45 and 55 values are now present in the parent node, that is, the index node.

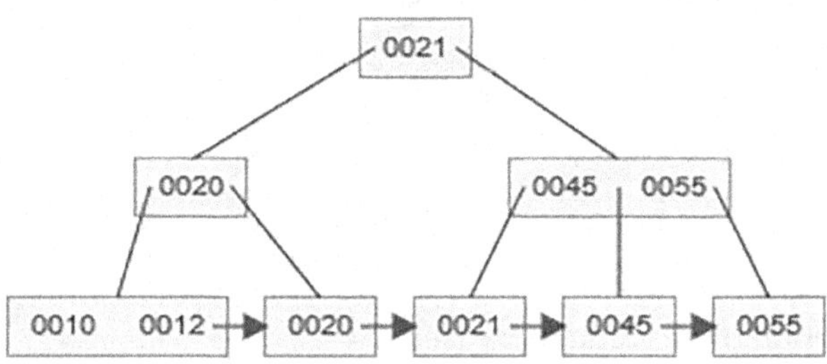

Deletion of 21 value from above B+ tree,

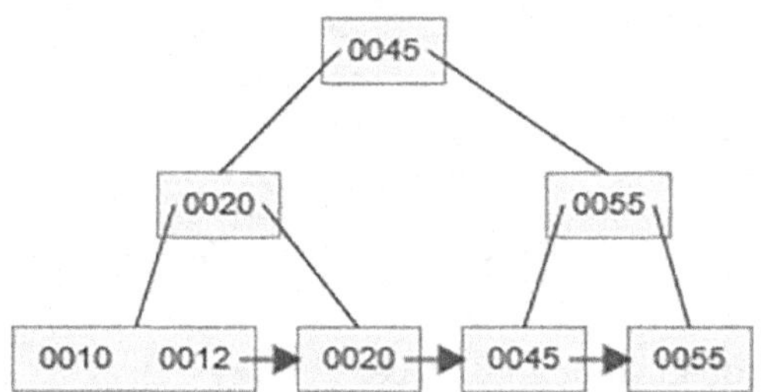

5.11.3.3 Search Operation in B+ Tree

In a B+ tree, the search operation is the same as that of BST. In a BST, the search process initiates from the root node, and each time we make a two-way decision: either we go to the left sub-tree or we go to the right sub-tree. In a B+ tree also, the search method starts from the root node, but every time we make an n-way decision where n is the total number of children that node has. In a B-tree, the search operation is carried out with O (log n) time complexity. The search process is carried out as follows:

Step 1: Start.

Step 2: First, read the search element from the end user.

Step 3: Then compare the search element with a first key value of the root node in the tree.

Step 4: If both are equal, then display "Given node is found" and terminate the function.

Step 5: If both are not matching, then check whether a search element is smaller or larger than that key value.

Step 6: If the search element is smaller, then continue the search process in the left sub-tree.

Step 7: If the search element is larger, then compare with the next key value in the same node and repeat steps 3, 4, 5 and 6 until we find an exact match or comparison completed with a last key value in a leaf node.

Step 8: If we accomplished with last key value in a leaf node, then display "Element is not found" and wind up the function.

Step 9: Stop.

5.11.4 Uses of B+ Tree

1. For the external sorting purpose, B+ tree is used.
2. B+ trees are used to provide indexed sequential file organization.
3. B+ tree is used in a database management system.

5.12 Interview Questions

1. What are different binary tree traversal techniques?
2. Implement preorder traversal of binary tree using recursion?
3. Explain the difference between binary tree and BST with an example?
4. What are advantages and disadvantages of BST?
5. What is AVL Tree?
6. Why do we want to use BST?
7. Write algorithm for post-order traversing?
8. How do you insert a new element in a BST?
9. What is the minimum number of nodes that a binary tree can have?
10. Write a note on heaps?
11. Explain different ways of deleting of nodes in the BST.
12. Explain tree terminologies with suitable examples?
13. Explain different types of binary tree representation?
14. Write a note on breadth first traversal and depth first traversal?
15. Explain recursive algorithm for in-order traversal?
16. What are the different operations performed on BSTs?
17. Explain different applications of heaps?

18. What is B tree?
19. What is the difference between B Tree and B+ tree?
20. What are advantages and disadvantages of the AVL tree?
21. Define binary tree. Explain different types of binary trees.

5.13 Multiple Choice Questions

1. Assume that we have numbers between 1 and 1000 in a BST and want to search for the number 363. Which of the below ordering could not be the ordering of the node reviewed?

   A. 2, 252, 401, 398, 330, 344, 397, 363

   B. 924, 220, 911, 244, 898, 258, 362, 363

   C. 925, 202, 911, 240, 912, 245, 258, 363

   D. 2, 399, 387, 219, 266, 382, 381, 278, 363

   Answer: C

2. In a complete binary tree or perfect binary tree, every inner node has exactly two children. If there are 21 leaf nodes or exterior nodes in the tree, how many internal nodes are there in the perfect binary tree?

   A. 19

   B. 20

   C. 22

   D. 18

   Answer: B
   Explanation:
   In a complete binary tree or perfect binary tree having n interior nodes, then there are exactly (n+1) leaf or exterior nodes.

3. Which type of traversal of BST produces the value in ascending sorted order?

   A. Preorder traversal

   B. Post-order traversal

   C. In-order traversal

   D. All of the above

   Answer: C

4. A 2-3 is a type of tree which is having the following properties,
   a. 2-3 tree's all internal nodes have either two or three children.
   b. 2-3 tree's all paths from root to leaves have the same length
   c. Then the number of internal nodes of a 2-3 tree having nine leaves could be
      A. 4 or 8
      B. 5 or 9
      C. 6 or 7
      D. 7 or 4

   Answer: D

5. If a node has two children that node we want to delete from a BST, then that deleted node must be replaced by its
   A. In-order successor
   B. In-order predecessor
   C. Post-order successor
   D. Post-order predecessor

   Answer: A

6. A BST is formed from the given sequence 6, 9, 1, 2, 7, 14, 12, 3, 8 and 18. The minimum number of nodes required to be added into this tree to form an extended binary tree is?
   A. 3
   B. 6
   C. 8
   D. 11

   Answer: D

7. In a full binary tree, every internal node has exactly two children. A full binary tree with 2n+1 nodes contains,
   A. 2n leaf nodes
   B. n+1 internal nodes
   C. n+1 exterior nodes
   D. n-1 internal nodes

   Answer: C

8. If n numbers are to be sorted in ascending order in O (n log n) time, which of the following tree can be used

A. Binary tree
B. Binary search tree
C. Max-heap
D. Min-heap

Answer: D

9. If n elements are sorted in a balanced BST. What would be the asymptotic complexity to search a key in the tree?
A. A O(1)
B. O(logn)
C. O(n)
D. O(nlogn)

Answer: B

10. In which of the following tree, parent node has a key value greater than or equal to the key value of both of its children?
A. Binary search tree
B. Threaded binary tree
C. Complete binary tree
D. Max-heap

Answer: D

11. A complete binary tree T has n leaf nodes. The number of nodes of degree 2 in T is
A. $\log_2 n$
B. n-1
C. n
D. 2^n

Answer: B

12. A BST is generated by inserting in order the following integers: 50, 15, 62, 5, 20, 58, 91, 3, 8, 37, 60 and 24.
The number of the node in the left sub-tree and right sub-tree of the root, respectively, is
A. (4, 7)
B. **(7, 4)**
C. (8, 3)
D. (3, 8)

Answer: B

13. The number of edges or links from the root node to the deepest leaf is called _________ of the tree.
    - A. Height
    - B. Depth
    - C. Length
    - D. Width

    Answer: A

14. In a complete binary tree or a proper binary tree or a strictly binary tree if the number of internal nodes is N, then the number of leaves or exterior nodes, E are?
    - A. E = 2*N -1
    - B. **E = N + 1**
    - C. E = N – 1
    - D. E = N – 2N

    Answer: B

15. Which of the following statements is false or true?
    1. A full binary tree or a proper binary tree or a strictly binary tree is a binary tree in which each node other than the leaves has one or two children.
    2. All complete binary trees are strictly binary trees, but all strictly binary trees are not complete binary trees.
    - A. Statement 1 is false.
    - B. Statement 2 is false.
    - C. Statements 1 and 2 both are false.
    - D. **Statements 1 and 2 both are true.**

    Answer: (D)

16. Which type of binary tree produces the value in ascending sorted order when it is traversed?
    - A. Binary search tree
    - B. Complete binary tree
    - C. Strictly binary tree
    - D. Almost complete binary tree

    Answer: A

17. Which of the following statements is false or true?
    1. If a tree is dominated by the left child node or right child node, it is said to be a skewed binary tree.

2. Forest is the disjoint union of trees.
   A. Statement 1 is false
   B. Statement 2 is false
   C. Statements 1 and 2 both are false
   D. **Statements 1 and 2 both are true**

   Answer: (D)

18. Which of the following statements is false or true?
    1. An AVL tree is also a BST, but it is a balanced tree.
    2. Every AVL Tree is a BST, and all the BSTs are also AVL trees.
       A. Statement 1 is false
       B. **Statement 2 is false**
       C. Statements 1 and 2 both are false
       D. Statements 1 and 2 both are true

    Answer: (B)

19. Which of the following statements is false or true?
    1. B tree is a binary tree.
    2. B+ tree is used to provide indexed sequential file organization.
       A. **Statement 1 is false**
       B. Statement 2 is false
       C. Statements 1 and 2 both are false
       D. Statements 1 and 2 both are true

    Answer: (A)

20. Which of the following statements is false or true?
    1. In a B+ tree, the search operation is different than that of BST.
    2. Internal sorting purpose B+ tree is used.
       A. Statement 1 is false
       B. Statement 2 is false
       C. **Statements 1 and 2 both are false**
       D. Statements 1 and 2 both are true

    Answer: (C)

21. Which of the following statements is false or true with respect to a B+ tree?
    1. In a B+ tree, all the leaf nodes must be at the same level.

2. In a B+ tree, all leaves have been connected to form a linked list of keys in sequential order.
   A. Statement 1 is false
   B. Statement 2 is false
   C. Statements 1 and 2 both are false
   D. **Statements 1 and 2 both are true**

Answer: (C)

6

Nonlinear Data Structures: Graph

6.1 Concepts and Terminology of Graph

6.1.1 Introduction to Graph

As we know, in the tree structure, the main restriction is that each tree has one root node. However, if we remove that restriction, we end up with a more complex data structure called a graph. In the graph, there is not a root node at all.

6.1.2 Definition of Graph

A graph is defined as a set of nodes or vertices and a set of edges or arcs that connect the two vertices.

A set of vertices is described by listing the vertices as in a set. For example, V = {k, l, m, n} and the set of edges is determined as a sequence of edges. For example, E = {(k, l), (m, n), (k, n), (l, m)}

The graph is commonly defined as follows:

G = (V, E)

where V (G) is a finite, non-empty set of vertices of a graph, G. E (G) is a set of pairs of vertices or arcs each representing an edge of a graph.

6.1.3 Types of Graph

6.1.3.1 Directed, Undirected and Mixed Graph

Directed graphs are also called digraphs. A directed graph or digraph is a collection of nodes connected by edges, where the edges have a direction linked with them.

An undirected graph is a type of graph, in which a set of vertices or nodes are connected, and all the edges are bidirectional.

In directed graphs, the directions are shown on the edges (Figure 6.1).

DOI: 10.1201/9781003105800-6

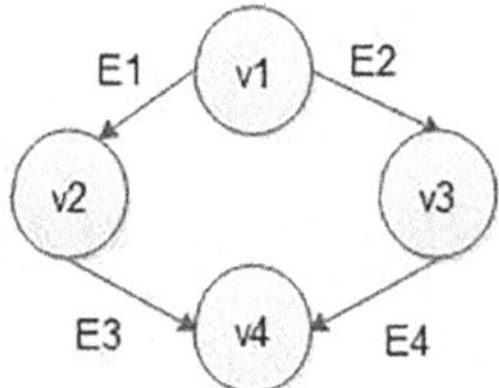

FIGURE 6.1
Directed graph.

As shown in Figure 6.1., the edges between vertices are ordered. In this type of graph, edge E_1 is in between vertices v1 and v2. The v1 is called head and v2 is called tail. We can say, E_1 is set of (V_1, V_2) and not of (V_2, V_1).

Similarly, in an undirected graph, edges are not ordered (Figure 6.2).

In this type of graph, that is an undirected graph, the edge E_1 is set of (V_1, V_2) or (V_2, V_1).

Consider the following directed graph (Figure 6.3):

V(G) = {A, B, C, D}

E(G) = {(A,A),(B,A),(B,D),(D,C),(A,B),(B,C),(A,C)}

In Figure 6.3, edge (A, B) and edge (B, A) are two different edges through both the edges that connect vertex A and vertex B. Some edges may connect

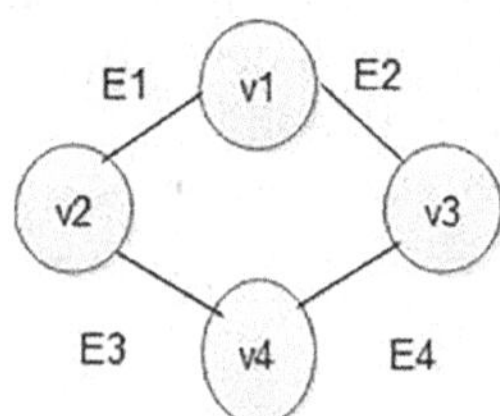

FIGURE 6.2
An undirected graph.

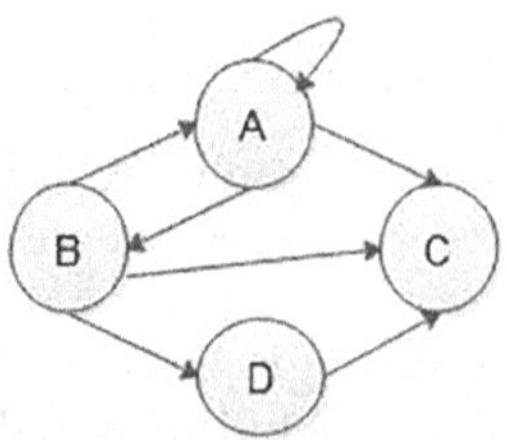

FIGURE 6.3
Directed graph.

one node with itself. These edges are called loops. For example, edge (A, A) is a loop.

The degree of a vertex is the number of edges incident to it. In-degree of a vertex is the number of edges pointing to that vertex and out-degree of a vertex is the number of edges pointing from that vertex. For example, the out-degree of node B in Figure 6.3 is 3, and its in-degree is 1 and degree of node B is 4.

Note: A tree is a special case of a directed graph.

A graph need not be a tree, but a tree must be a graph. In a graph, a node that is not adjacent to any other node is called an isolated node (Figure 6.4).

For example, node D in a graph is an isolated node shown in Figure 6.4. The degree of an isolated node is zero.

A graph with undirected and directed edges is said to be a mixed graph as shown in Figure 6.5.

6.1.3.2 Null Graph

A graph containing only an isolated node is called a null graph. A null graph contains an empty set of edges (Figure 6.6).

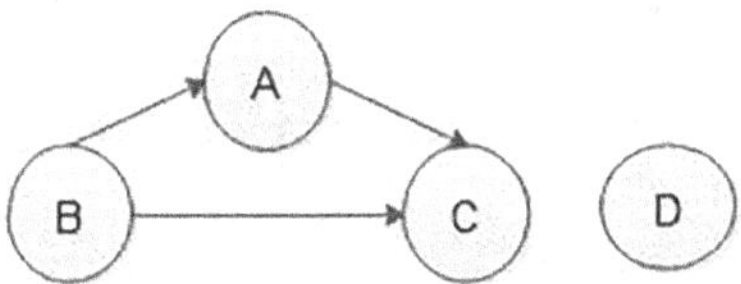

FIGURE 6.4
Graph containing isolated node D.

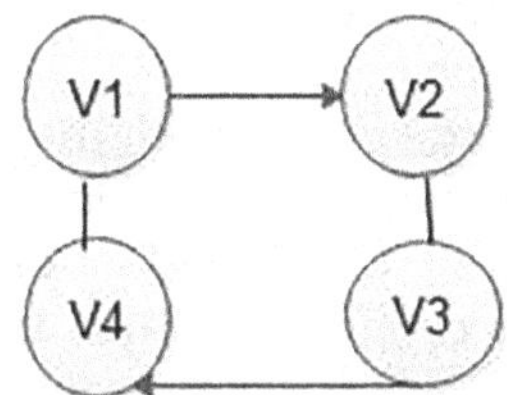

FIGURE 6.5
Mixed graph.

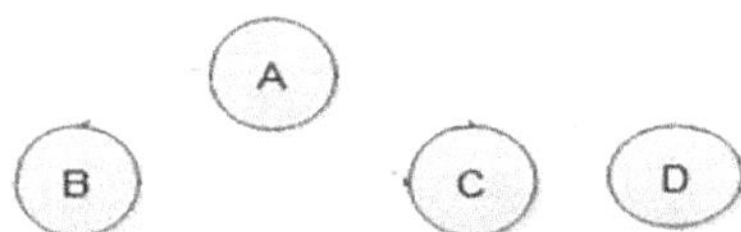

FIGURE 6.6
Null graph.

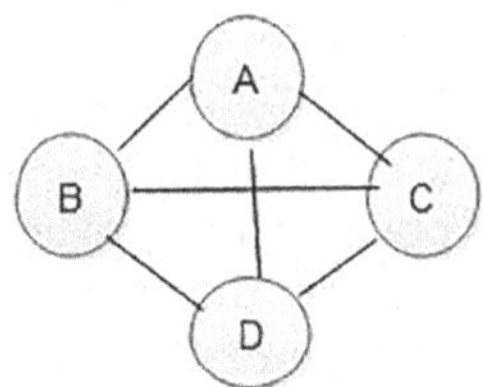

FIGURE 6.7
Complete graph.

6.1.3.3 Complete Graph

If an undirected simple graph of 'n' vertices consists of n (n-1)/2 number of edges, then it is called a complete graph. The complete graph does not contain any loop edge. Here each node is connected. If the graph contains four nodes, then each node's degree is three in a complete graph as shown in Figure 6.7.

6.1.3.4 Strongly Connected Graph

A digraph is said to be strongly connected if for every pair of vertex there exists a path, that is, in such a graph path between V1 and V2 exist, then there is a path from V2 to V1 is also present. For example, in Figure 6.8, a graph is a strongly connected graph.

6.1.3.5 Unilaterally Connected and Weakly Connected Digraph

A simple digraph is said to be **unilaterally connected** if for any pair of nodes of a graph at least one of the nodes of a pair is reachable from the other node.

The concept of **"strongly connected"** and **"weakly connected"** graphs are concerned with directed graphs.

If for any pair of nodes of the graph, both the nodes of the pair are reachable from one another, then the graph is called **strongly connected**.

If the digraph is treated as an **undirected graph (means directions are not considered),** and when it is connected graph, there exists a path from any pair of vertex to another and only then the graph is called **weakly connected graph**.

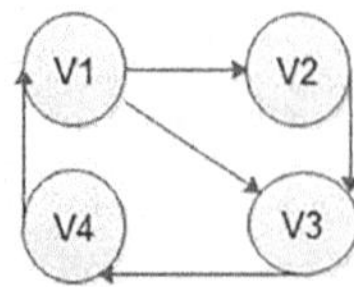

FIGURE 6.8
Strongly connected graph.

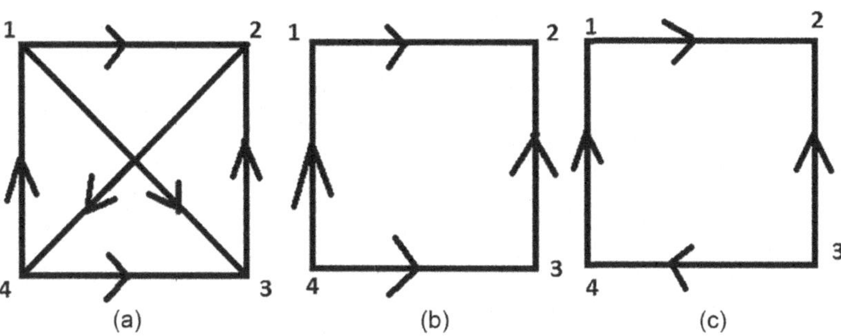

FIGURE 6.9
Unilaterally connected, strongly connected, and weakly connected digraph. (a) Strongly connected graph. (b) Weakly connected graph but not unilaterally connected. (c) Unilaterally connected but not strongly connected.

A unilaterally connected digraph is a graph in which every pair of node their exists either a path from node v1 to v2 or v2 to v1 for each pair of node as shown in Figure 6.9b.

While a unilaterally connected digraph is **weakly connected**, but a weakly connected digraph is not necessarily **unilaterally connected.**

A strongly connected digraph is, at the same time, unilaterally and weakly connected.

6.1.3.6 Simple Graph

A graph G is called a simple graph if the graph does not contain any loop or parallel edges. The graph shown in Figure 6.10 is simple.

6.1.3.7 Multigraph

A graph G is called a multigraph if a graph contains loop or parallel edges. The graph shown in Figure 6.11 is a multigraph.

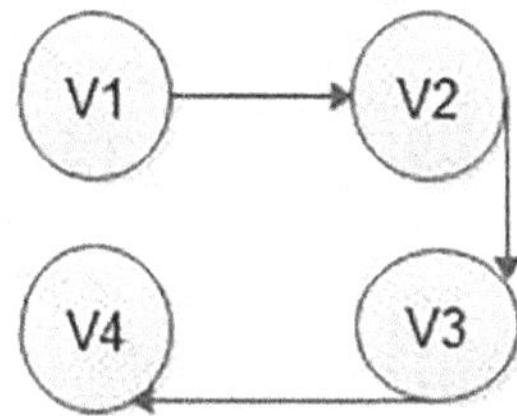

FIGURE 6.10
Simple graph.

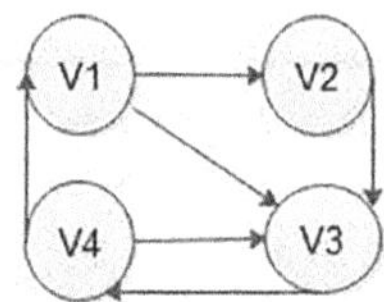

FIGURE 6.11
Multigraph.

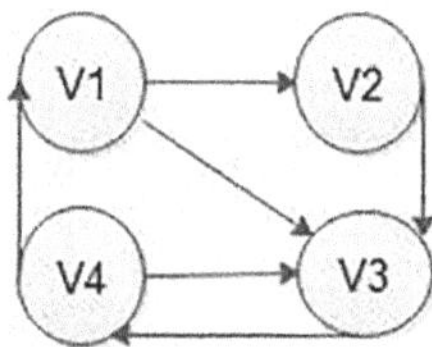

FIGURE 6.12
Cyclic graph.

6.1.3.8 Cyclic Graph

A graph with at least one cycle is known as a cyclic graph.

In Figure 6.12, the graph contains v1-v2-v3-v4-v1, v1-v3-v4-v1 and v3-v4-v3 cycles, so the above graph is called a cyclic graph.

6.1.3.9 Acyclic Graph or Non-cyclic Graph

A graph **with no** cycles is called an acyclic graph or a non-cyclic graph.

In Figure 6.13., the graph does not have any cycles. Hence, it is a non-cyclic graph or an acyclic graph.

6.1.3.10 Bipartite Graph

A simple graph G = (V, E) with vertex or nodes are partition into set V1 and V2 where V = {V1, V2} and $V1 \cap V2 = \phi$ is called a bipartite graph if every edge of E joins a vertex in V1 to a vertex in V2.

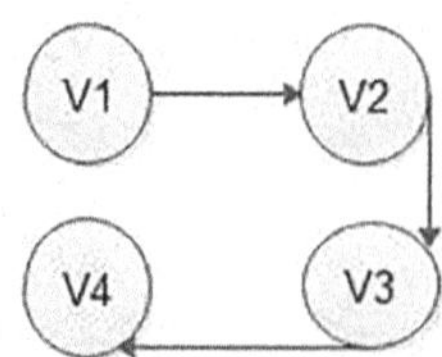

FIGURE 6.13
Acyclic graph.

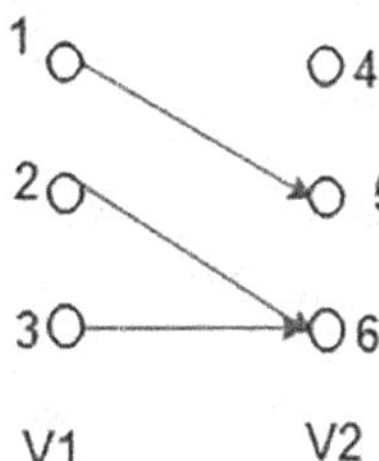

FIGURE 6.14
Bipartite graph.

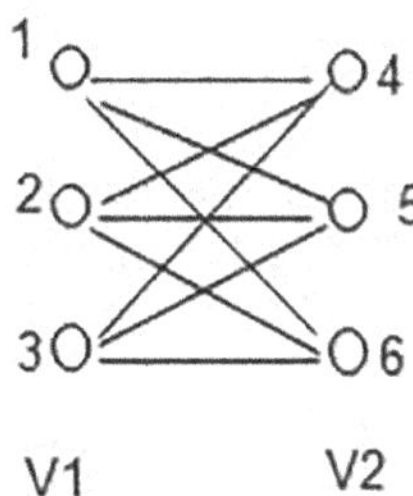

FIGURE 6.15
Complete bipartite graph.

In general, a bipartite graph has two sets of vertices, let us say, V1 and V2, and if an edge is drawn, it should connect any vertex in set V1 to any vertex in set V2. The graph shown in Figure 6.14 is bipartite.

6.1.3.11 Complete Bipartite Graph

A complete bipartite graph is a simple graph in which the vertices can be partitioned into two disjoint sets V1 and V2 such that each vertex in V1 is adjacent to each vertex in V2. A complete bipartite graph is shown in Figure 6.15.

Here, a bipartite graph is a simple graph where every vertex of the first set V1 is connected to every vertex in the second set V2.

6.2 Representation of Graph Using Adjacency Matrix and Adjacency List

There are two commonly used graph representation techniques:

1. Adjacency matrix representation
2. Adjacency lists representation

6.2.1 Adjacency Matrix Representation

6.2.1.1 Definition of Adjacency Matrix

Consider a graph G with a set of vertices V(G) and a set of edges E(G). If there are N nodes in V(G), for N >= 1, the graph G may be represented by an adjacency matrix which is a table with N rows and N columns where

A(i, j) = 1 if and only if there is an edge (V_i, V_j) in E(G)

0 otherwise

If there is an edge connecting V_i and V_j in E (G), the value in the [i, j] position in the table is 1, otherwise, it is 0.

In this representation, the graph may be represented using a matrix of a total number of vertices per total number of vertices. This means if a graph with five vertices can be represented using a matrix of size 5 × 5. Within this matrix, both rows and columns are vertices. This matrix is filled with either 1 or 0. Here, 1 represents that there is an edge from row vertex to column vertex, and 0 describes that there is no edge from row vertex to a column vertex.

6.2.1.2 Example of Adjacency Matrix

See Figures 6.16 and 6.17.

If the relation between two nodes is unordered, the graph is called an unordered or undirected graph (Figures 6.18 and 6.19).

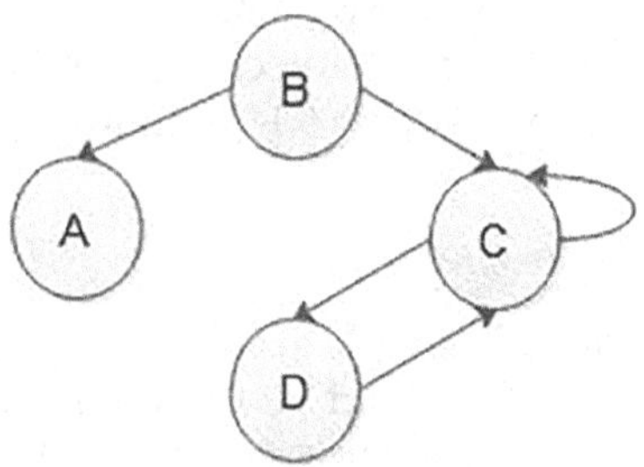

FIGURE 6.16
Directed graph, G.

	A	B	C	D
A	0	0	0	0
B	1	0	1	0
C	0	0	1	1
D	0	0	1	0

FIGURE 6.17
Adjacency matrix of directed graph G.

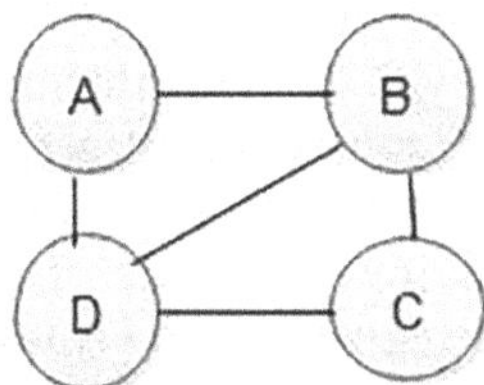

FIGURE 6.18
Undirected graph, G.

	A	B	C	D
A	0	1	0	1
B	1	0	1	1
C	0	1	0	1
D	1	1	1	0

FIGURE 6.19
Adjacency matrix of undirected graph G.

6.2.1.3 Algorithm of Creating or Representing Graph Using Adjacency Matrix

The algorithm for the creation of the graph using adjacency matrix will be as follows:

1. Declare an array of M [size] [size] which will store the graph, where the size = number of maximum nodes which we have stored in the graph.
2. Enter how many nodes you want in a graph.
3. Enter edges of graph by two vertices each say V_i, V_j indicates some edge.
4. If the graph is directed set M [i] [j] = 1. If the graph is undirected set M [i] [j] = 1 and M [j] [i] = 1 as well.
5. When all edges in the desired graph are entered, print the graph M [i] [j].

6.2.2 Adjacency List Representation

6.2.2.1 Basics of Adjacency List Representation

The use of the adjacency matrix to represent a graph is inappropriate due to its static implementation. To use this representation, we must know in advance the number of nodes that a graph has in order to set up storage. The solution to this problem is to use a linked structure, which makes allocations and de-allocations from an available pool. We will represent the graph using adjacency lists. This adjacency list stores information about only those edges that exist. The adjacency list contains a directory and a set of linked lists. This representation is also known as **node directory representation**. The directory contains one entry for each node of the graph. Each entry in the directory points to a linked list that represents a node that is connected to that node. Directory represent nodes and the linked list represent the edges.

Each node of the linked list has three fields:

1. Node identifier,
2. Link to next field,
3. An optional weight field contains the weight of the edge.

An undirected graph can also be stored using an adjacency list; however, each edge will be represented twice, once in each direction.

In general, an undirected graph of order N, which is the number of nodes N, with E edges requires N entries in the directory and 2*E linked list entries, whereas a directed graph will require N entries in the directory and E linked list entries.

6.2.2.2 Example of Adjacency List Representation for Directed Graph

Figure 6.22 shows the node directory representation of the directed graph as shown in Figure 6.20.

The out-degree of a node in a directed or undirected graph can be determined by counting the number of entries in its linked list. Here, in Figure 6.21, node 2 having three nodes in its linked list, that is, node 2 has out-degree 3. However, node 2 does not present in the linked list means node 2 having zero in-degree. Similarly, count node 4 in the linked list part, which is two in number, so node 4 has in-degree two.

However, it is difficult to determine the in-degree of node, but we can easily obtain the out-degree (Figure 6.22).

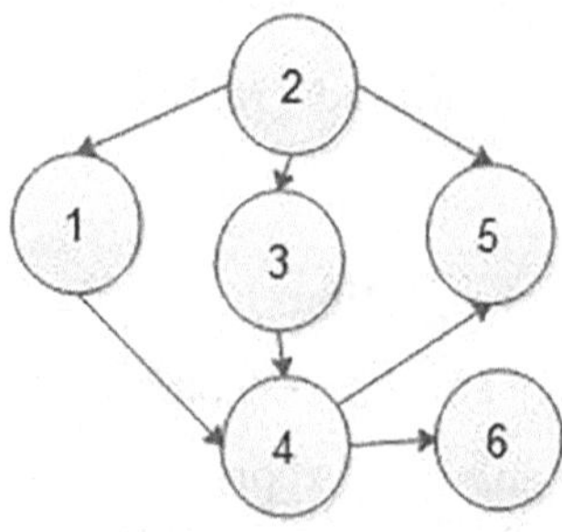

FIGURE 6.20
A directed graph.

FIGURE 6.21
Sample node.

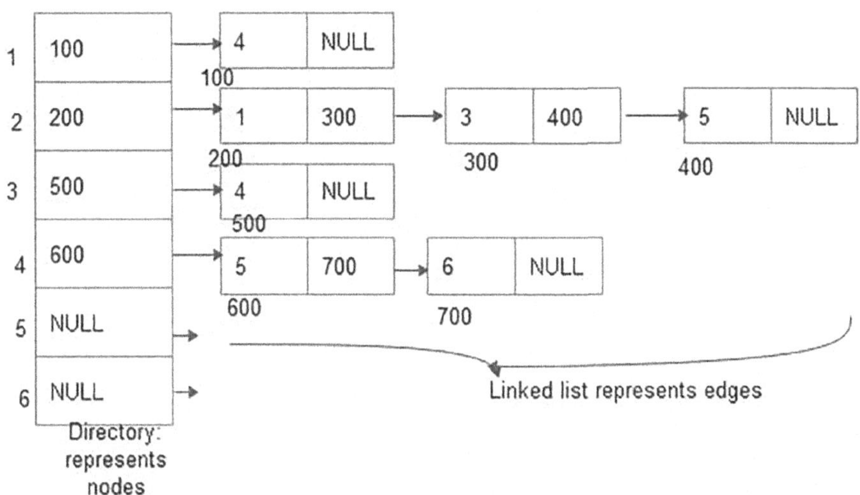

FIGURE 6.22
Node directory representation for directed graph.

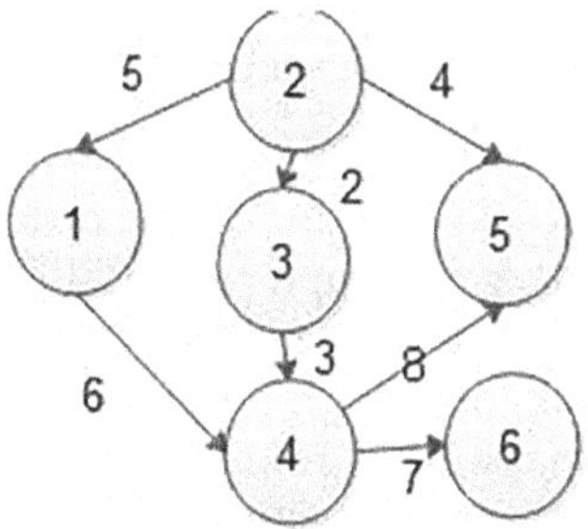

FIGURE 6.23
Weighted graph.

6.2.2.3 Example of Adjacency List Representation for Weighted Graph

Figure 6.23 shows a weighted graph (Figure 6.24), and Figure 6.25 shows its node directory representation for a weighted graph.

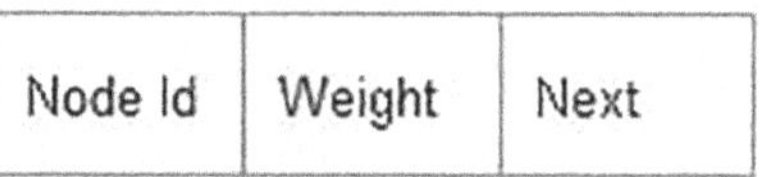

FIGURE 6.24
Sample node.

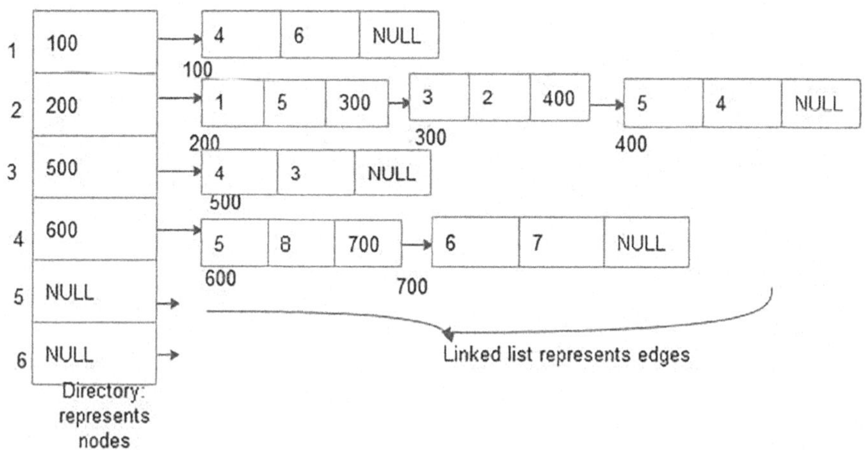

FIGURE 6.25
Node directory representation for a weighted graph.

6.3 Graph traversal Techniques (Breadth First Search and Depth First Search)

Graph traversal is a procedure used to search for a vertex or node in a graph. The graph traversal is also used to decide the order of vertices be visited in the search process only once. A graph traversal finds the edges to be used in the search process without creating loops that means using graph traversal, we visit all vertices of the graph only at one time without getting into a looping path. The majority of graph problems involve the traversal of a graph. Traversal of a graph means visiting each node exactly once.

There are two types of graph traversal techniques, and they are as follows:

1. Breadth first search (BFS)
2. Depth first search (DFS)

6.3.1 BFS Traversal of a Graph

6.3.1.1 Basics of BFS

In the graph, say vertex V_1 in a graph will be visited first, then all vertices adjacent to V_1 will be traversed suppose adjacent to V_1 are (V_2, V_3) then again from V_2 adjacent vertices will be printed. This process will be continued for all the vertices to be encountered. To keep track of all vertices and their adjacent vertices, we will make use of an array for the visited nodes. The nodes that get visited are set to 1. Thus, we can keep track of visited nodes.

6.3.1.2 Algorithm of BFS Traversal of a Graph

```
// Array visited [ ] is initialized to 0.
// q is queue-type data structure.
// BFS traversal on graph G is carried out beginning at
   vertex V.
void  BFS( int v)
{
  q: a queue type variable
  Initialize q;
  visited[ v] = 1;     //mark v as visited
  add vertex v to queue q;
  while( q is not empty )
  {
  Print node v;
  v  → delete an element from the queue ;
  for all vertices w adjacent from v
  {
       if ( ! visited[ w] )
       {
       visited[ w] = 1;
       add the vertex w to queue q;
       } // completion of if statement
 } // completion of for loop
} // completion of while loop
} // completion of BFS function
```

6.3.1.3 Example of BFS

Find BFS traversal for following graph G (Figure 6.26).

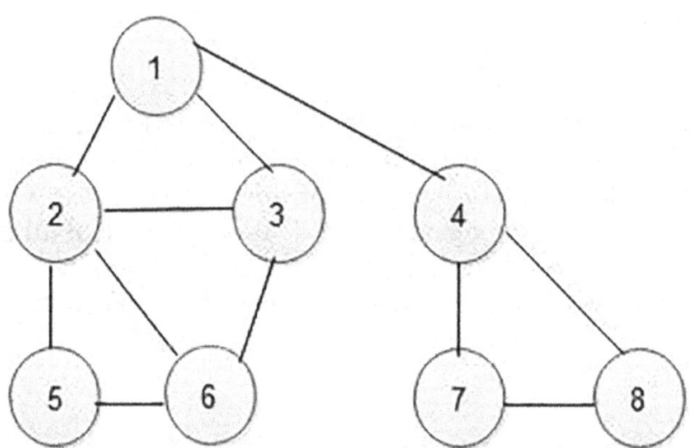

FIGURE 6.26
Given graph G.

Solution:

Step 1:

Start from BFS (1),

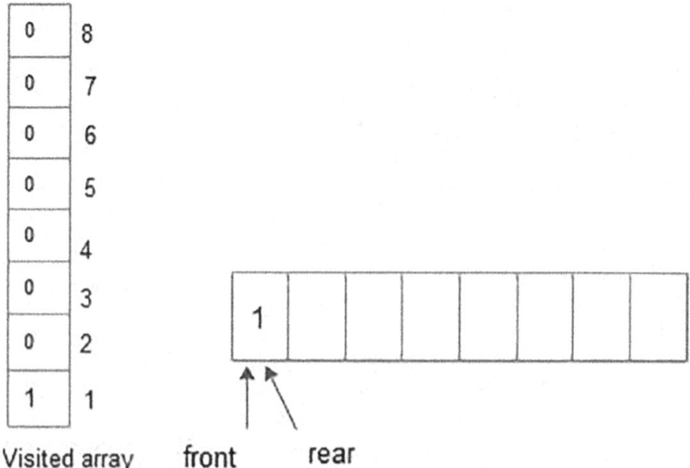

Print value 1,

Delete node 1 from the queue and insert not visited adjacent nodes to the queue.

Step 2:

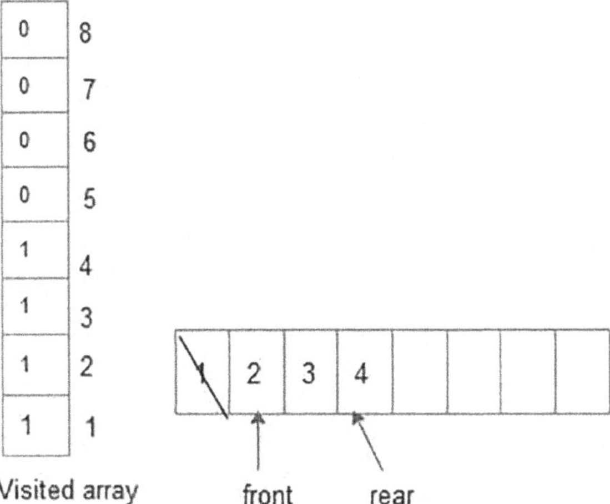

Print value 2,

Delete node 2 from the queue and insert not visited adjacent nodes of 2 into the queue.

Step 3:

Adjacent nodes of 2 are {1, 5, 3, 6} but 1 and 3 are already visited so discard it and add 5 and 6 nodes into the queue.

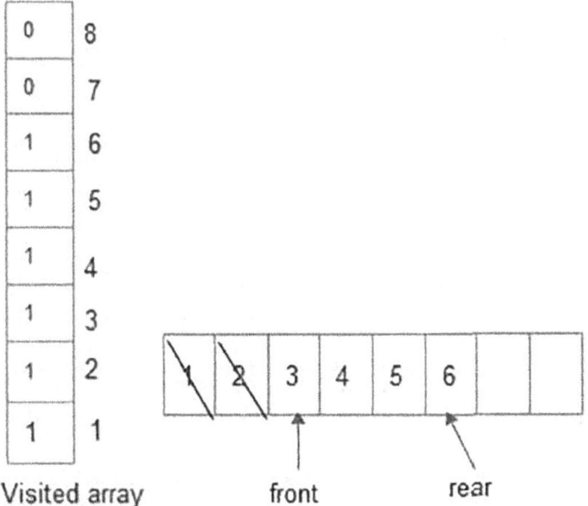

Print value 3,

Delete node 3 and insert non-visited adjacent node to the queue, then

Step 4:

Adjacent nodes of node 3 are {1, 2, 6}, but all are already visited so discard it, and no node is added into the queue.

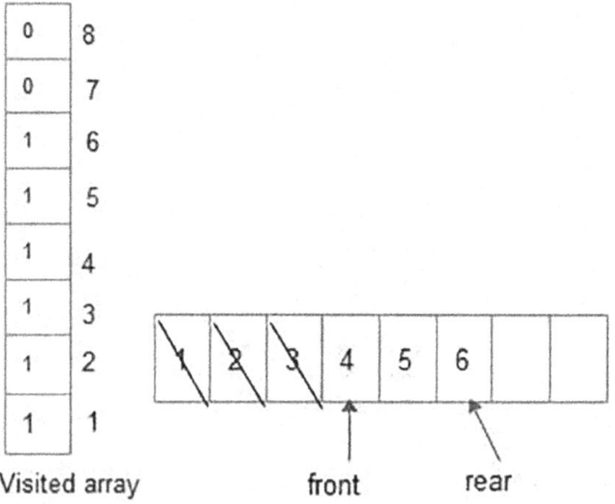

Print value 4,

Delete node 4 and insert non-visited adjacent nodes to the queue, then

Step 5:

Adjacent nodes of node 4 are {1, 7, 8}, but 1 is already visited so discard it and insert 7 and 8 to the queue.

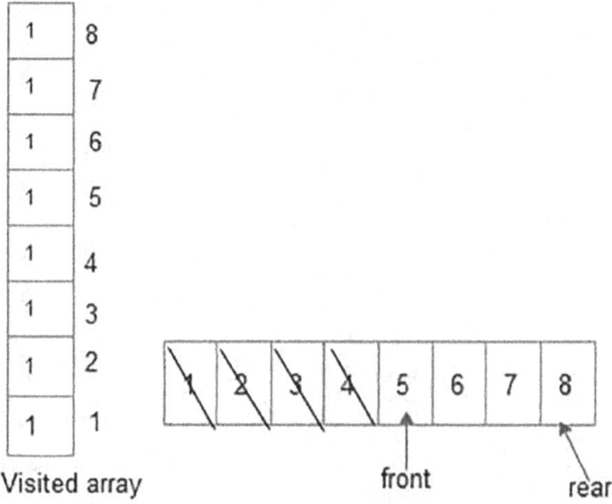

Print value 5,

Delete node 5 and insert non-visited adjacent nodes to the queue, then

Step 6:

Adjacent nodes of node 5 are {6, 2} already visited.

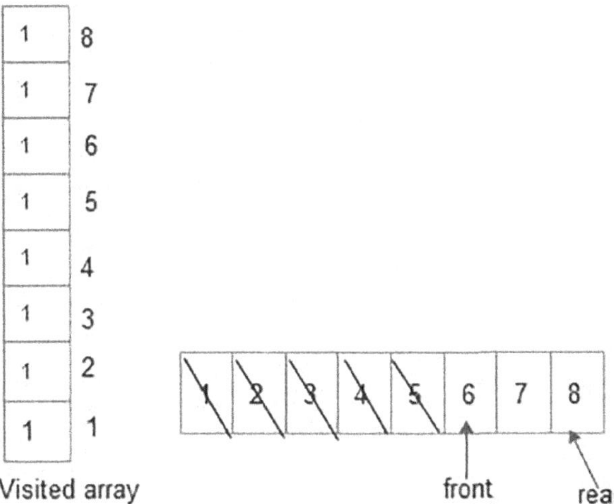

Print value 6,

Delete node 6 and insert non-visited adjacent nodes to the queue, then

Step 7:

Adjacent nodes of node 6 are {5, 2, 3} already visited.

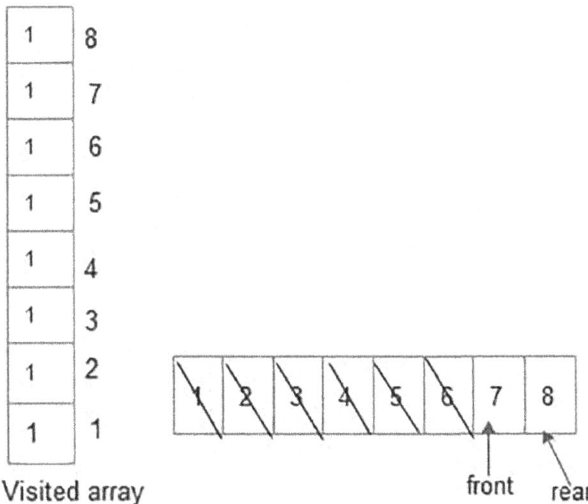

Print value 7,

Delete node 7 and insert non-visited adjacent node to the queue, then

Step 8:

Adjacent nodes of node 7 are {4, 8} already visited.

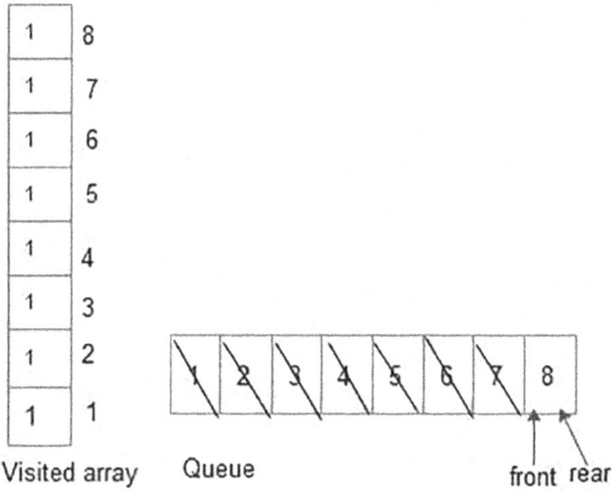

Print value 8,

Delete node 8 and insert non-visited adjacent nodes to the queue, then

Step 9:

Adjacent nodes of node 8 are {7, 4} already visited.

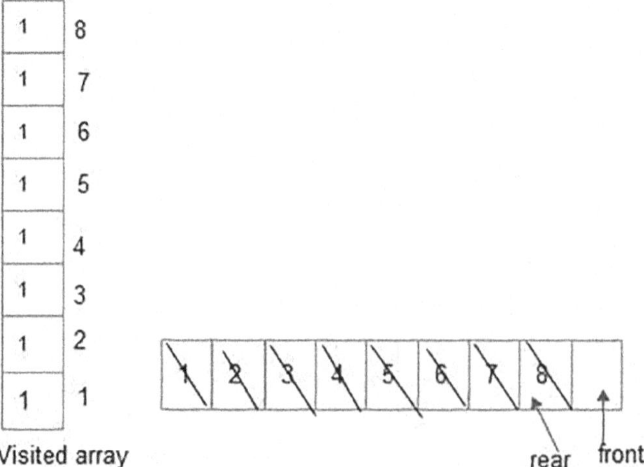

Queue is now empty because front>rear or front index cross to the rear index, this condition shows that queue is empty, then stop the BFS algorithm.

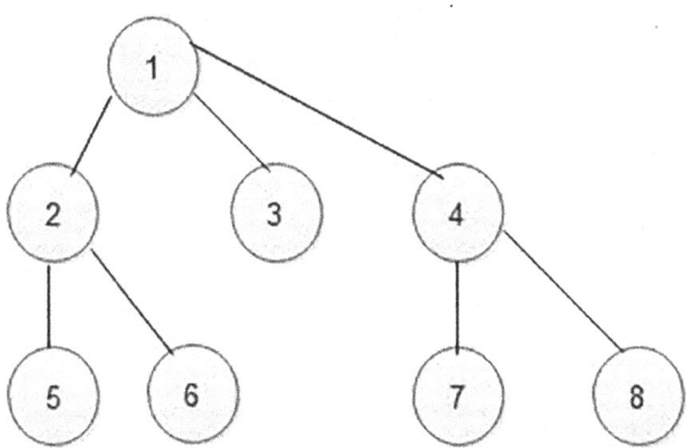

FIGURE 6.27
BFS spanning tree of Figure 6.25.

We get the following trace as follows:

1	1	1	1	2	2	4	4
1	2	3	4	5	6	7	8

By using this trace, we will find the BFS spanning tree, which is as follows:

Before that, spanning tree is a subset of Graph G, which has all the nodes or vertices covered with the minimum possible number of edges. Hence, a spanning tree does not have cycles or loops, and it cannot be disconnected (Figure 6.27).

Nodes of the tree are printed in order of their level.

Finally, the nodes are visited by using BFS is as follows:

1 – 2 – 3 – 4 – 5 – 6 – 7 – 8.

6.3.1.4 Implementation of Graph Traversing Using the BFS Technique

```
#include<stdio.h>
#include<stdlib.h>
#define MAX 100
#define initial 1
#define waiting 2
#define visited 3
int n;
int adj[MAX][MAX];
int state[MAX];
void create_graph(); //Create a graph using adjacency matrix
```

```
void BF_Traversal();  // traverse a graph using BFS technique
void BFS(int v);        // Breadth first search function

int queue[MAX], front = -1,rear = -1;
void insert_queue(int vertex);
int delete_queue();
int isEmpty_queue();

// main function definition
int main()
{
create_graph();
BF_Traversal();
return 0;
}

void BF_Traversal()
{
int v;
for(v=0; v<n; v++)
state[v] = initial;
printf("Enter start Vertex for BFS: \n");
scanf("%d", &v);
BFS(v);
}
// pass node v as the first node for traversing graph using
BFS traversing technique
void BFS (int v)
{
int i;
insert_queue(v);
state[v] = waiting;
//Continue the while loop till queue is not empty
while( !isEmpty_queue() )
{
v = delete_queue( );
printf("%d ",v);
state[v] = visited;
//check all adjacent nodes of node v and check whether the
adjacent nodes are visited or not
for(i=0; i<n; i++)
{
if(adj[v][i] == 1 && state[i] == initial)
{
insert_queue(i);
state[i] = waiting;
} //if completed
} // for loop completed
} // while loop completed
printf("\n");
}
```

```
//insert vertex or node in the queue
void insert_queue (int vertex)
{
if(rear == MAX-1)
{
printf("Queue is full\n");
}
else
{
if(front == -1)
{
front = 0;
}
rear = rear+1;
queue[rear] = vertex ;
}
}
//Queue is empty or not function
int isEmpty_queue ()
{
if(front == -1 || front > rear)
return 1;
else
return 0;
}
//delete node from the queue
int delete_queue ()
{
int delete_item;
if (front == -1 || front > rear)
{
printf ("Queue is empty \n");
exit(1);
}
delete_item = queue[front];
front = front+1;
return delete_item;
}
//create adjacency graph and initialize it
void create_graph()
{
int count, max_edge, origin, destin;
printf("Enter number of vertices : ");
scanf("%d", &n);
max_edge = n*(n-1);
for(count=1; count<=max_edge; count++)
{
printf("Enter edge %d( -1 -1 to quit ) : ",count);  //exit
from for loop enter -1 -1
scanf("%d %d", &origin, &destin);
```

```
if((origin == -1) && (destin == -1))
break;

if(origin>=n || destin>=n || origin<0 || destin<0)//Check node
numbers are in between 0 and n-1
{
printf("Invalid edge!\n");
count--;
}
else
{
adj[origin][destin] = 1;  //Otherwise insert the edge into the
adjacency array
}
} // for loop completed
}
```

Output:

```
Enter number of vertices : 4
Enter edge 1( -1 -1 to quit ) : 0 1
Enter edge 2( -1 -1 to quit ) : 0 3
Enter edge 3( -1 -1 to quit ) : 1 2
Enter edge 4( -1 -1 to quit ) : -1 -1
Enter start Vertex for BFS:
0
0 1 3 2
```

6.3.1.5 Time Complexity of BFS

If the graph is implemented using an **adjacency matrix**, that is, an array of size V × V, then, for each node, you have to traverse an entire row of length V in the matrix to discover all its outgoing edges. Therefore, the time complexity of BFS is **T (n) = O(n^2) where n is the number of vertices in the graph.**

If your graph is implemented using **adjacency lists or node directory representation**, wherein each node maintains a list of all its adjacent edges, then, for each node, you could discover all its neighbors by traversing its adjacency list just once in linear time. Therefore, the time complexity of BFS is **T (n) = O (V + E)**, where V is the number of vertices in the graph and E is the number of edges in the graph.

6.3.1.6 Advantages of BFS

1. In this procedure in any way, it will find the goal.
2. It finds the minimal solution in the case of multiple paths.

6.3.1.7 Disadvantages of BFS

1. BFS consumes large memory space. Its time complexity is more.
2. It has long pathways when all paths to a destination are on approximately the same search depth.

6.3.1.8 Applications of BFS

1. Finding the shortest path from source to other vertices in an unweighted graph.
2. Finding the quick and efficient solution of a puzzle such as Rubik's cube by applying BFS on the state space.
3. Creating bipartite graphs.

6.3.2 DFS Traversal of a Graph

6.3.2.1 Basics of DFS

In the DFS traversal of a graph, we follow the path as deeply as we can go. When there is no adjacent vertex present, then we travel backward and look for the not-visited vertex. We will maintain a visited array to mark all the visited vertices. A node already reported as visited should not be selected for the traverse. Marking of visited vertices can be performed using a global array visited[ ]. The visited array[ ] is set to zero.

6.3.2.2 Non-recursive DFS Algorithm

1. Push node V1 on the stack.
2. Print node V1 value.
3. Mark node V1 value in the visited array as 1.
4. Find all adjacent nodes of node v1.
5. Repeat step 4 until all adjacent node ends.
6. If node v1 does not have any adjacent node, then pop node V1 from the stack.
7. Repeat steps 1–6 until the stack is empty.

6.3.2.3 Recursive DFS Algorithm

Step 1: Start

Step 2: n - number of nodes in graph G.

Step 3: Initialize visited [ ] to 0

```
for(i=0; i<n; i++)
visited [i] = 0
```

Step 4: Void DFS (vertex i)

```
{
print i;
visited [i] = 1;
for each w adjacent to i
if (! visited [w])
DFS (w);
}
```

Step 5: stop.

6.3.2.4 Explanation of Logic for Depth First Traversal (DFS)

In DFS, the basic data structure for storing the adjacent nodes is **stack**. In our program, we have used a recursive call to the DFS function. When a recursive call is invoked, a push operation is performed. When we exit from the loop, a pop operation will be performed (Figure 6.28).

Step 1: Start with vertex 0, print it so '0' node is printed. Mark 0 node as visited.

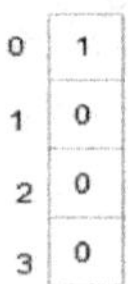

Visited array

Step 2: Find adjacent vertex to 0, there are two adjacent vertices 1 and 2, out of that choose one say node 1 is chosen if it is not visited, call DFS (1), that is, 1 will get inserted into the stack, mark node 1 as visited in the visited array.

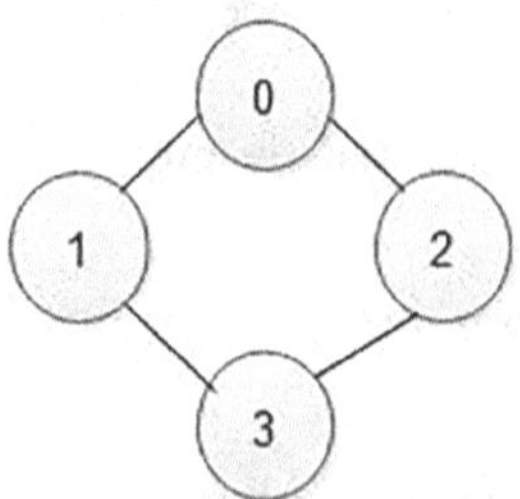

FIGURE 6.28
An undirected graph.

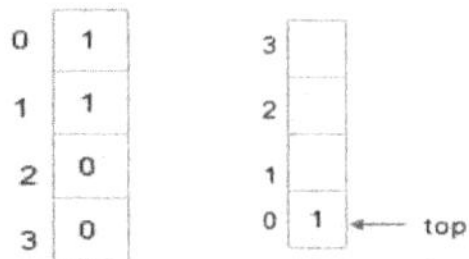

After exiting loop 1 will be popped and print 1.

Step 3: Find adjacent to 1 that is vertex 0 or 3 but 0 is already visited, so remaining node 3 is chosen if it is not visited and call DFS (3), that is, 3 will get pushed onto stack, mark it as visited in a visited array.

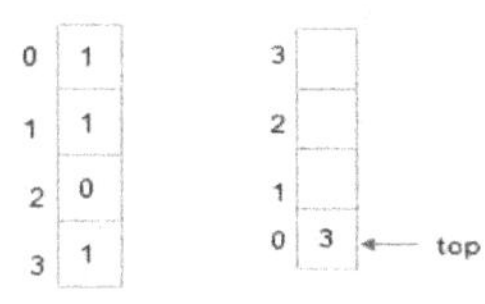

After existing, loop 3 will be popped and print 3.

Step 4: Find adjacent to 3 that is vertex 2 if it is not visited call DFS (2), that is, 2 will be pushed onto the stack and mark node 2 as visited in a visited array.

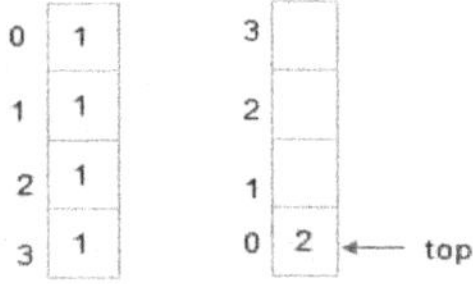

After existing, loop 2 will be popped and print 2.

Since all nodes are visited once then stop the procedure.

So the output of DFS is 0 1 3 2.

6.3.2.5 Implementation of Graph Traversing Using the DFS Technique

```
#include<stdio.h>
void DFS( int );   //DFS function declaration
int G[10][10], visited[10], n;  //n is number of vertices and
                                 graph is stored in array G[10][10]
                                 //visited[10] is visited array
//main function definition
int main()
{
int i, j;
```

```
printf("Enter number of vertices: \n");
scanf("%d",&n);
//read the adjacency matrix, enter 1 if edge is present and 0
if edge is not present
printf("Enter adjacency matrix of the graph: \n");
for(i=0;i<n;i++)
{
        for(j=0;j<n;j++)
        {
        scanf("%d",&G[i][j]);
        }
}
//visited array is initialized to zero
for(i=0;i<n;i++)
{
        visited[i]=0;
}
//call DFS with 0 node as start node
printf("DFS traversal of a graph : ");
DFS(0);
return 0;;
}
//Recursive DFS function definition
void DFS (int i)
{
int j;
printf("\t %d",i);
visited[i]=1;
for(j=0;j<n;j++)
{
        if( !visited[j] && G[i][j] == 1)
        {
        DFS(j);
        }
}
}
```

Output:

```
Enter number of vertices:
4
Enter adjacency matrix of the graph:
0 1 1 0
1 0 0 1
1 0 0 1
0 1 1 0
DFS traversal of a graph :      0       1       3       2
```

6.3.2.6 Time Complexity of DFS

If graph is implemented using an **adjacency matrix**, that is, an array of size V × V, then, for each node, you have to traverse an entire row of length V in the matrix to discover all its outgoing edges. Therefore, the time complexity of DFS is **T (n) = O (n^2) where n is the number of vertices in the graph.**

If your graph is implemented using **adjacency lists or node directory representation**, wherein each node maintains a list of all its adjacent edges, then, for each node, you could discover all its neighbors by traversing its adjacency list just once in linear time. Therefore, the time complexity of DFS is **T (n) = O (V + E), where V is the number of vertices in the graph and E is the number of edges in the graph.**

6.3.2.7 Advantages of DFS

1. DFS consumes very little memory space.
2. It will reach the goal node in less period than BFS if it traverses in the right path.

6.3.2.8 Disadvantages of DFS

1. Many states may keep reoccurring.
2. Sometimes the states may also enter into infinite loops.

6.3.2.9 Applications of DFS

1. For finding connected components of the graph.
2. For finding strongly connected components.
3. For solving maze problems.
4. For finding cycles in graphs and finding the largest cycle in graph.

6.4 Applications of Graph as Shortest Path Algorithm and Minimum Spanning Tree

6.4.1 Shortest Path Algorithm

Let G = (V, E) be a graph with n vertices.

The problem is to find out the shortest distance from the vertex to all other vertices of a graph.

6.4.1.1 *Dijkstra's Algorithm for Shortest Path*

Dijkstra's algorithm is also called **a single source shortest path algorithm.** It is based on a **greedy or optimal technique.** Greedy algorithm is the algorithm, which picks the best solution at the moment without regard for consequences. In this algorithm, we use

1. Visited array
2. Cost matrix
3. Distance matrix

6.4.1.2 *Dijkstra's Algorithm*

1. Start.
2. Create a cost matrix C [ ] [ ] from the adjacency matrix adj[ ] [ ].
   C [i] [j] is a cost of going from vertex i to vertex j.
   If there is no edge between vertices i and j, then C [i] [j] is set to be infinity.
3. Array visited [ ] is initialized to zero.

```
for (i=0; i<n; i++)
visited [i]=0;
```

4. If vertex 0 is the source vertex, then visited [0] is marked as 1.
5. Build the distance matrix by storing the cost of vertices from vertex number 0 to (n-1) from the source vertex 0.

```
for (i=1; i<n; i++)
      distance [i] = cost [0] [i] ;
```

   Initial distance of source vertex is initiated as 0. that is. distance [0].

6. `for (i=1; i<n; i++)`

   Choose a vertex w, such that distance [w] is minimum and visited [w] is 0.

   Mark visited [w] as 1.

   Recompute the shortest distance of rest vertices from the source. Only the vertices not

   marked as 1 in array visited [ ] should be considered for recalculation of distance.

   That is,

```
for each vertex v
if (visited[v]==0)
distance [v]=min (distance[v], distance[w] + cost[w][v])
```

7. Stop.

6.4.1.3 *Time Complexity*

$T(n) = O(n^2)$

6.4.1.4 *Example Showing Working of Dijkstra's Algorithm*

Show working of Dijkstra's algorithm on a graph given as below (Figure 6.29):

Solution:
Source vertex is taken as 0.
Cost matrix:

	0	1	2	3	4
0	∞	10	∞	30	100
1	10	∞	50	∞	∞
2	∞	50	∞	20	10
3	30	∞	20	∞	60
4	100	∞	10	60	∞

Distance matrix:

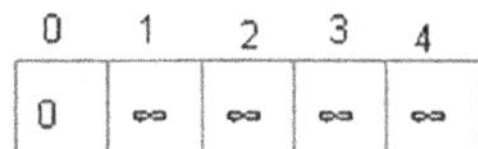

0	1	2	3	4
0	∞	∞	∞	∞

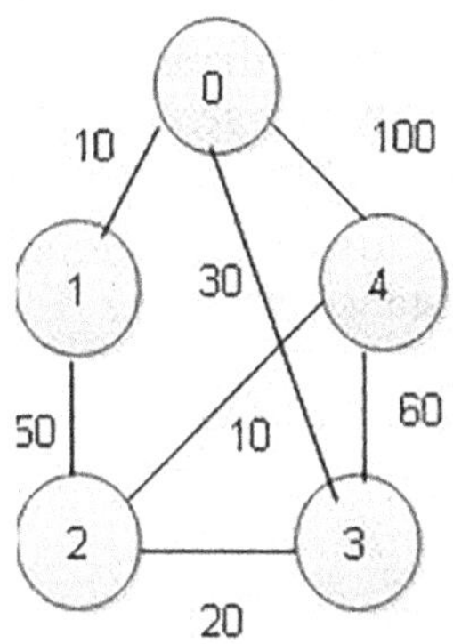

FIGURE 6.29
Given undirected weighted graph G.

Visited array:

0	1	2	3	4
0	0	0	0	0

First iteration:

1. Select vertex 0.
2. Mark visited [0] as 1.
3. Calculate distance of vertex 1 from 0 using cost matrix. Cost of vertex 1 from 0 is 10.
4. Cost of vertices 3 and 4 from vertex 0 is 30 and 100, respectively. Therefore, update the distance matrix.

0	1	2	3	4
0	10	∞	30	100

Distance matrix:

0	1	2	3	4
1	0	0	0	0

Visited array

Second iteration:

1. Select vertex 1.
2. Mark visited [1] as 1.
3. Re-adjust distances.
4. Cost of going to vertex 2 from the source vertex 0, through selected vertex 1 is given by, distance [1]+cost [1] [2] = 10 +50 = 60.
   Distance of 60 is better than existing distance of ∞ or infinity.
5. Cost of going to vertex 3 from the source vertex 0, through selected vertex 1 is given by, distance [1]+cost [1] [3] = 10+∞ = ∞
   which is worse than existing distance of 30.
6. Similarly, cost of going to vertex 4 through vertex 1 is ∞ which is worse than existing distance of 100.

0	1	2	3	4
0	10	60	30	100

Distance matrix:

0	1	2	3	4
1	1	0	0	0

Visited array:

Third iteration:
Vertex selected = 3
Cost of going to vertex 2 through 3 = dist[3]+cost[3] [2] = 30+20 = 50
Cost of going to vertex 4 through 3 = dist[3]+cost[3] [4] = 30+60 = 90
Distance of vertices 2 and 4 should be changed.

0	1	2	3	4
0	10	50	30	90

Distance matrix:

0	1	2	3	4
1	1	0	1	0

Visited array:
Fourth iteration:
Vertex selected = 2
Cost of going to vertex 4 through 2 = dist[2]+cost[2] [4] = 50+10 = 60
Distance of vertex 4 should be changed.

0	1	2	3	4
0	10	50	30	60

Distance matrix with final distances:

0	1	2	3	4
1	1	1	1	0

Visited array:

6.4.2 Minimum Spanning Tree

6.4.2.1 Basic Concept of Spanning Tree and Minimum Spanning Tree

A spanning tree is a minimal sub-graph of a given graph and they follow or satisfy all of the following three conditions:

1. A number of vertices in a spanning tree are equal to the number of vertices in a given graph G.
2. A minimal sub-graph or tree has a lesser number of edges than the given graph G.
3. Minimal spanning tree should be connected and there should not be any cycle.

Simply, a **spanning tree** is a subset of an undirected Graph that has all the vertices connected by a minimum number of edges.

If all the vertices are connected in a graph, then there exists at least one spanning tree. In a graph, there may exist more than one spanning tree. If graph G, contains 'n' vertices, then the spanning tree contains 'n' vertices and (n-1) edges.

6.4.2.2 Properties Spanning Tree

1. There is no cycle in a spanning tree.
2. Any vertex can be reached from any other vertex.

A **Minimum Spanning Tree (MST)** is a subset of the edges of a connected weighted undirected graph that connects all the vertices with the minimum possible total edge weight.

6.4.2.3 Example of Spanning Tree

Given Graph G =

Here in the spanning tree, all the vertices are visited as in graph G, but the edges are less than graph G (Figures 6.30 and 6.31).

Finding a spanning tree of a weighted graph G, having minimum cost can be calculated using **Greedy strategy**.

Feasible solution: Final graph must contain all vertices, and they must be connected with no cycle.

Optimal solution: It is a feasible solution with the minimum cost.

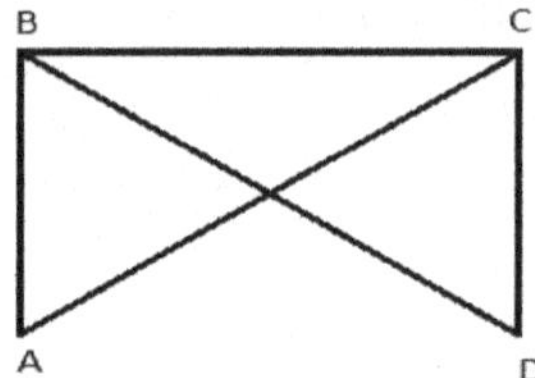

FIGURE 6.30
Undirected graph G.

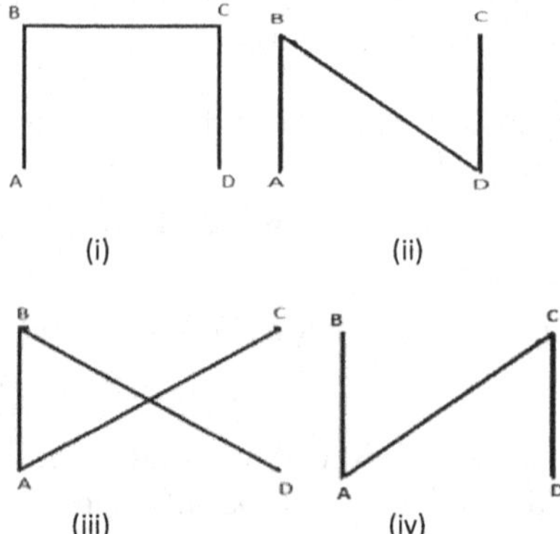

FIGURE 6.31
Given undirected graph G and its four spanning trees.

There are two types of algorithms to find MST:

1. Prime's algorithm
2. Kruskal's algorithm

6.4.2.4 Prime's Algorithm

Let G (V, E) is a connected weighted undirected graph with no loops and parallel edges. If graph G contains any parallel edge, then choose a less weight edge from the parallel edge and remove another one. Also, if any loop is there in the graph, then remove that loop also. Therefore, after removing the loop and the largest parallel edge if any from graph G. Now for graph G, we will find the MST using Prime's algorithm.

Prime's Algorithm

1. Label all the vertices as unchosen.
2. Let the tree T consist of 'n' nodes, initially with no edge. T is also called a solution vector.
3. Choose any arbitrary vertex 'u' and label it as chosen.
4. while (there is an unchosen vertex)

```
{
pick the lightest edge between any chosen 'u' and unchosen 'v'
label v as chosen T = T + <u, v>
if and only  <u, v> does not give rise to cycle in T
}
```

Now we will represent **Prime's algorithm in different ways** as follows:

Let G(V, E) is a connected weighted undirected graph with no loops and parallel edges.

1. Create two sets V and V′ such that V′ is empty initially.
   V contains all vertices of graph G.
   Select minimum weighted edge (i, j) from G and add i and j into set V′.
2. T is the solution vector that contains no edges initially, that is, it is an empty set of edges.
3. Repeat step 4 while V is not equal to V′.
4. Find all neighbors of all vertices which are in set V′ such that one endpoint of the neighboring edge is in V′ and another not in V′ or another is present in V. In addition, edge weight is minimum and not form a loop or cycle. That edge to add to the solution vector T.

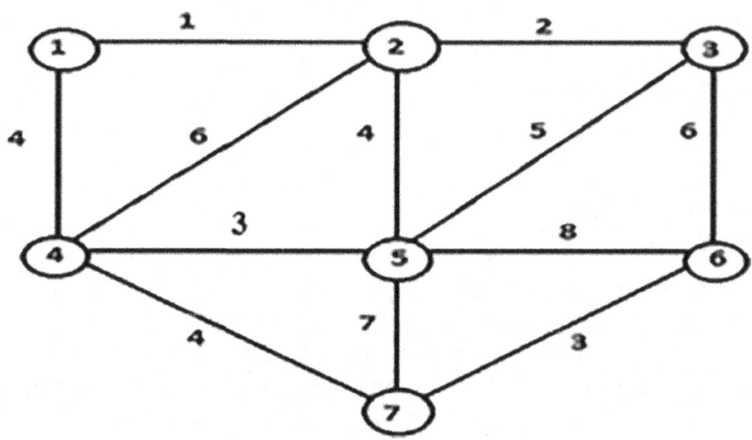

FIGURE 6.32
Given undirected weighted graph G.

5. Make the sum of all selected edge weights that gives us the minimum weighted spanning tree.
6. Stop.

Example: Calculate minimum cost-spanning tree using Prime's Algorithm (Figure 6.32).

Solution:

Adjacency Matrix :

1	2	3	4	5	6	7
∞	1	∞	4	∞	∞	∞
1	∞	2	6	4	∞	∞
∞	2	∞	∞	5	6	∞
4	6	∞	∞	3	∞	4
∞	4	5	3	∞	8	7
∞	∞	6	∞	8	∞	3
∞	∞	∞	4	7	3	∞

Step 1:

Set V = {1, 2, 3, 4, 5, 6, 7}
Set V′ = empty set initially
T = { } = empty set initially

Step 2: First iteration:

By choosing 1 as a starting node
V′ ={ 1, 2}
T ={ <1,2> }

Step 3: Second iteration:

<2, 3> edge having minimum weight and does not form a cycle so choose <2, 3> edge and insert into the solution vector T. Node 3 is inserted into the set V′.

V′ ={ 1, 2, 3}

T ={ <1, 2>, <2, 3> }

Step 4: Third iteration:

<1, 4> edge having minimum weight and does not form a cycle so choose <1, 4> edge and insert into the solution vector T. Node 4 is inserted into the set V′.

V′ ={ 1, 2, 3, 4}

T ={ <1, 2>, <2, 3>, <1, 4> }

Step 5: Fourth iteration:

<4, 5> edge having minimum weight and does not form a cycle so choose <4, 5> edge and insert into the solution vector T. Node 5 is inserted into the set V′.

V′ ={ 1, 2, 3, 4, 5}

T ={ <1, 2>, <2, 3>, <1, 4>, <4, 5> }

Step 6: Fifth iteration:

<4, 7> edge having minimum weight and does not form a cycle so choose <4, 7> edge and insert into the solution vector T. Node 7 is inserted into the set V′.

V′ ={ 1, 2, 3, 4, 5, 7}

T ={ <1, 2>, <2, 3>, <1, 4>, <4, 5>, <4, 7> }

Step 7: Sixth iteration:

<7, 6> edge having minimum weight and does not form a cycle so choose <7, 6> edge and insert into the solution vector T. Node 6 is inserted into the set V′.

V′ ={ 1, 2, 3, 4, 5, 7, 6}

T ={ <1, 2>, <2, 3>, <1, 4>, <4, 5>, <4, 7>, <7, 6> }

Step 8: Seventh iteration:

Now V′ = V, so terminate the loop.

Step 9: Add weights of all edges in solution vector T as,

1+2+4+3+4+3 =17

Step 10: Stop the algorithm.

The above algorithm is also explained in the following way (Figure 6.33):

1 **Select Node 1:**

a. $1 - 2 \rightarrow 1$ ✓

b. $1 - 4 \rightarrow 4$ ✓

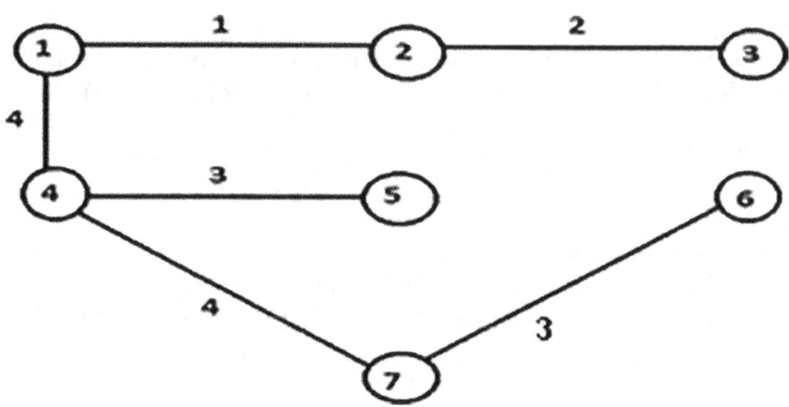

FIGURE 6.33
Minimum cost-spanning tree for graph G.

2 Select Node 2:

a. 2 – 1 → 1 ×
b. 2 – 3 → 2 ✓
c. 2 – 4 → 6
d. 2 – 5 → 4 ×

3. Select Node 3:

a. 3 – 2 → 2 ×
b. 3 – 5 → 5 ×
c. 3 – 6 → 6

4. Select Node 4:

a. 4 – 1 → 4 ×
b. 4 – 2 → 6
c. 4 – 5 → 3 ✓
d. 4 – 7 → 4 ✓

5. Select Node 5:

a. 5 – 2 → 4 ×
b. 5 – 3 → 5 ×
c. 5 – 4 → 3 ×
d. 5 – 6 → 8
e. 5 – 7 → 7

6. **Select Node 7:**
   a. $7-4 \rightarrow 4$ ×
   b. $7-5 \rightarrow 7$
   c. $7-6 \rightarrow 3$ ✓

7. **Select Node 6:**
   a. $6-3 \rightarrow 6$
   b. $6-5 \rightarrow 8$
   c. $6-7 \rightarrow 3$ ×

Minimum weight = 1+2+4+3+4+3 = 17

Another minimum cost-spanning tree for graph G can be as follows (Figure 6.34):

Minimum weight = 1+2+3+4+4+3 = 17

Disadvantages of Prime's Algorithm:

1. The list of edges has to be searched from the beginning as a new edge is added.
2. If there is more than one edge with the same weight, then all possible spanning trees are required to be found for minimal tree.

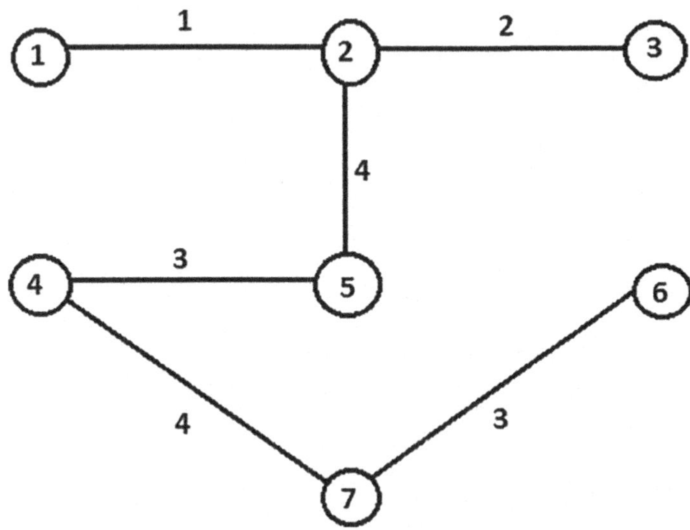

FIGURE 6.34
Minimum cost-spanning tree for graph G.

Complexity Analysis of Prime's Algorithm:

1. Complexity analysis is based on an algorithm that uses the cost matrix as a data structure and works on it. Since the matrix has to be scanned repeatedly the time complexity of the algorithm is **O (n^2).**
2. Prime's Algorithm does not use the Greedy Method fully because at the first step it gives the freedom to choose any vertex.

6.4.2.5 Kruskal's Algorithm

Kruskal's Algorithm uses the **Greedy approach** as it goes on choosing the lightest edges and then choose the next lightest edge and so on.

All the edges are sorted in **ascending order** while choosing the next edge. Care is to be taken that it should not give rise to **a cycle.**

Let G(V, E) is a connected weighted undirected graph with no loops and parallel edges.

1. T = null; T is a solution vector.
2. V_s = Set of Vertices
3. Construct a priority queue containing all edges in E in ascending order.
4. For each V ∈ V_s add {V} to V_s.
5. while ($|V_s| > 1$) // **if V_s set contains only one element then terminate the loop.**

```
        {
        choose <V, W> an edge ∈ Q (Priority Queue) with least
        weight.
        delete <V, W> from Q
        if (<V, W> are in different sets W1 and W2 in Vs)
        {

        replace W1 and W2 in Vs by W1 U W2
        add <V, W> to T
        }

        if (<V, W> are in the same set)
        {
        it suggests a cycle
        discard <V, W>
        }
    } // While loop complete
```

Example: Find out the minimum cost of a spanning tree by using Kruskal's algorithm (Figure 6.35).

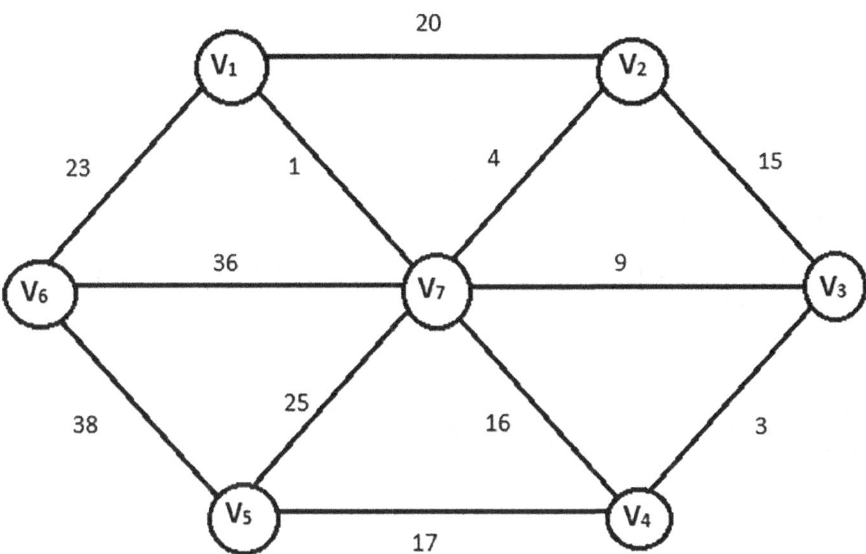

FIGURE 6.35
Given undirected weighted graph G.

Solution:

Step 1:

T = null

V_s = null

E = {<V_1, V_2>, <V_2, V_3>....} **//set of all edges**

T is a solution vector, initially null with no edges.

V_s is the set of vertices and at the initial step it is null.

Step 2: Arrange all the edges in ascending order according to their weights.

When you select any edge then add it to the set of vertices V_s and the Tree, T is solution vector.

Care must be taken that there should not be any cycle.

Priority Queue:

$V_1 - V_7 \rightarrow 1$ ✓
$V_3 - V_4 \rightarrow 3$ ✓
$V_2 - V_7 \rightarrow 4$ ✓
$V_3 - V_7 \rightarrow 9$ ✓
$V_2 - V_3 \rightarrow 15$ × **// creates loop so discard**
$V_4 - V_7 \rightarrow 16$ × **// creates loop so discard**
$V_4 - V_5 \rightarrow 17$ ✓
$V_1 - V_2 \rightarrow 20$ × **// creates loop so discard**

$V_1 - V_6 \rightarrow 23$ ✓
$V_5 - V_7 \rightarrow 25$
$V_6 - V_7 \rightarrow 36$
$V_5 - V_6 \rightarrow 38$
$V_s = \{\{V_1\}, \{V_2\}, \{V_3\}, \{V_4\}, \{V_5\}, \{V_6\}, \{V_7\}\}$

This is the set of vertices at the initial stage.

Step 3: We select edge $V_1 - V_7$ because it has minimum weight. Add it to V_s and T.
$V_s = \{\{V_1, V_7\}, \{V_2\}, \{V_3\}, \{V_4\}, \{V_5\}, \{V_6\}\}$
$T = \{<V_1, V_7>\}$

Step 4: Next select $V_3 - V_4$ with weight 3.
$V_s = \{\{V_1, V_7\}, \{V_2\}, \{V_3, V_4\}, \{V_5\}, \{V_6\}\}$
$T = \{<V_1, V_7>, <V_3, V_4>\}$

Step 5: Next edge is $V_2 - V_7$ with weight 4.
$V_s = \{\{V_1, V_7, V_2\}, \{V_3, V_4\}, \{V_5\}, \{V_6\}\}$
$T = \{<V_1, V_7>, <V_3, V_4>, <V_2, V_7>\}$

Step 6: Next edge is $V_3 - V_7$ with weight 9.
$V_s = \{\{V_1, V_7, V_2, V_3, V_4\}, \{V_5\}, \{V_6\}\}$
$T = \{<V_1, V_7>, <V_3, V_4>, <V_2, V_7>, <V_3, V_7>\}$

Step 7:
The next edge is $V_2 - V_3$ with weight 15. We discard this edge because it results in cycle. After all, V_2 and V_3 are in the same set in V_s.
Therefore, the next edge is $V_4 - V_7$ with weight 16, we discard this edge also because it results in cycle. V4 and V_7 are in the same set in V_s.
Therefore, the next edge is $V_4 - V_5$ with weight 17. Select it.
$V_s = \{\{V_1, V_7, V_2, V_3, V_4, V_5\}, \{V_6\}\}$
$T = \{<V_1, V_7>, <V_3, V_4>, <V_2, V_7>, <V_3, V_7>, <V_4, V_5>\}$

Step 8:
The next edge is $V_1 - V_2$ with weight 20. We discard this edge because it results in cycle. V_1 and V_2 are in the same set in V_s.
The next edge is $V_1 - V_6$ with weight 23. Select it.
$V_s = \{\{V_1, V_7, V_2, V_3, V_4, V_5, V_6\}\}$
$T = \{<V_1, V_7>, <V_3, V_4>, <V_2, V_7>, <V_3, V_7>, <V_4, V_5>, <V_1, V_6>\}$

Step 9: So all vertices of the given graph are visited. Stop this process. We get tree which is **minimum cost-spanning tree** as follows (Figure 6.36):
Minimum cost of spanning tree= $23+1+4+9+3+17 = 57$

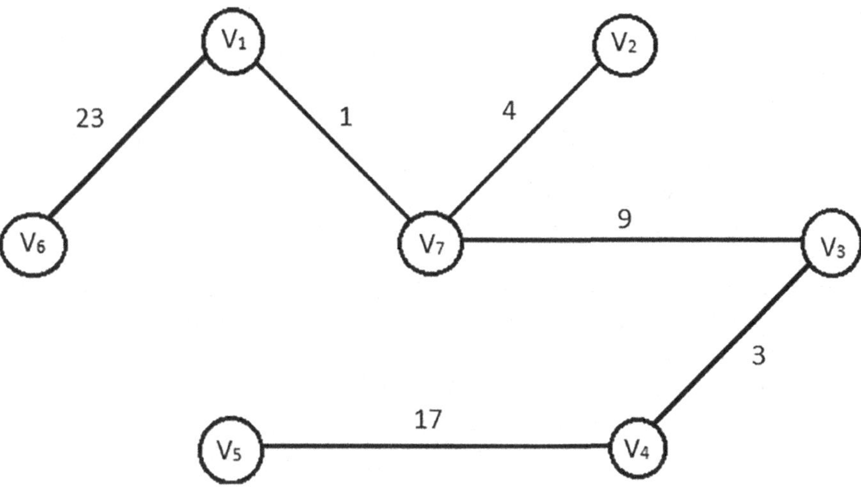

FIGURE 6.36
Minimum cost-spanning tree for graph G.

Note: The above tree satisfies all the conditions to become a spanning tree and mainly it uses **greedy strategy** because we choose or select the edge with minimum weight as the first edge instead of choosing any random edge like in Prime's Algorithm.

Example: Find minimum cost-spanning tree by using Kruskal's algorithm (Figure 6.37).

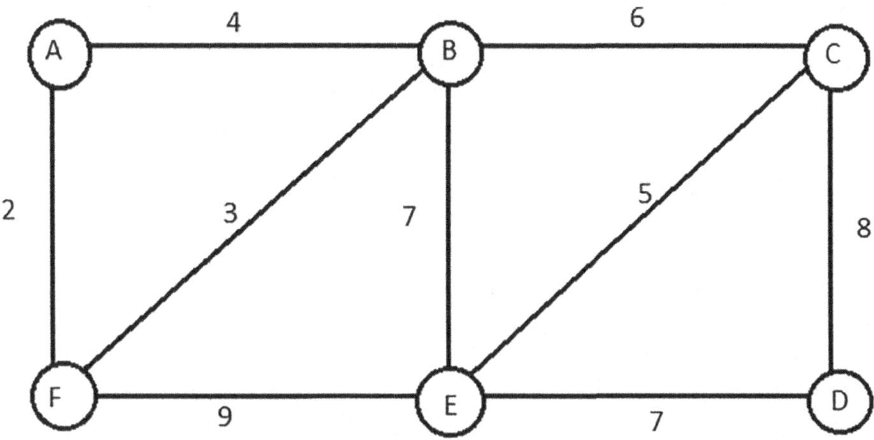

FIGURE 6.37
Given undirected weighted graph G.

Solution:

Step 1:

T = null

V_s = null

E = {<V_1, V_2>, <V_2, V_3>….} **//set of all edges**

T is a solution vector, initially null with no edges.

V_s is the set of vertices and at the initial step it is null.

Step 2: Arrange all the edges in ascending order according to their weights.

When you select any edge, then add it to the set of vertices V_s and the Tree, T is the solution vector.

Care must be taken that there should not be any cycle.

Priority Queue:

A – F → 2 ✓
B – F → 3 ✓
A – B → 4 × **// creates loop so discard**
C – E → 5 ✓
B – C → 6 ✓
B – E → 7 × **// creates loop so discard**
D – E → 7 ✓
C – D → 8
E – F → 9
V_s = {{A}, {B}, {C}, {D}, {E}, {F}}

This is the set of vertices at the initial stage.

Step 3: We select edge A – F because it has minimum weight. Add it to V_s and T.

T = {<A, F>}

V_S = {{A, F}, {B}, {C}, {D}, {E}}

Step 4: We select edge B – F because it has minimum weight. Add it to V_s and T.

T = {<A, F>, <B, F>}

V_S = {{A, F, B}, {C}, {D}, {E}}

Step 5:

The next edge is A – B with weight 4. We discard this edge because it results in cycle. A and B are in the same set in V_s.

Step 6:

Therefore, the next edge is C – E with weight 5. Select it.

T = {<A, F>, <B, F>, <C, E>}

V_S = {{A, F, B}, {D}, {C, E}}

Step 7:

Therefore, the next edge is B – C with weight 6, select it.

T = {<A, F>, <B, F>, <C, E>, <B, C>}

V_S = {{A, F, B, C, E}, {D}}

Step 8:

The next edge is B – E with weight 7. We discard this edge because it results in cycle. In addition, B and E are in the same set in V_s.

Step 9:

Therefore, next edge is D – E with weight 7, select it.

T = {<A, F>, <B, F>, <C, E>, <B, C>, <D, E>}

V_S = {{A, F, B, C, E, D}}

Step 10: So all vertices of a given graph are visited. Stop this process. We get tree that is **minimum cost-spanning tree** as follows (Figure 6.38):

Minimum cost = 2+3+5+6+7 = 23

Time complexity of Kruskal's Algorithm:

1. Adding all vertices to V_s would require the time complexity O (n).
2. In the while loop, we are supposed to take a union of two sets, which can be done in O (n).
3. Whenever an edge is chosen and deleted the queue, it shows the corresponding status. These operations require the complexity O (1).

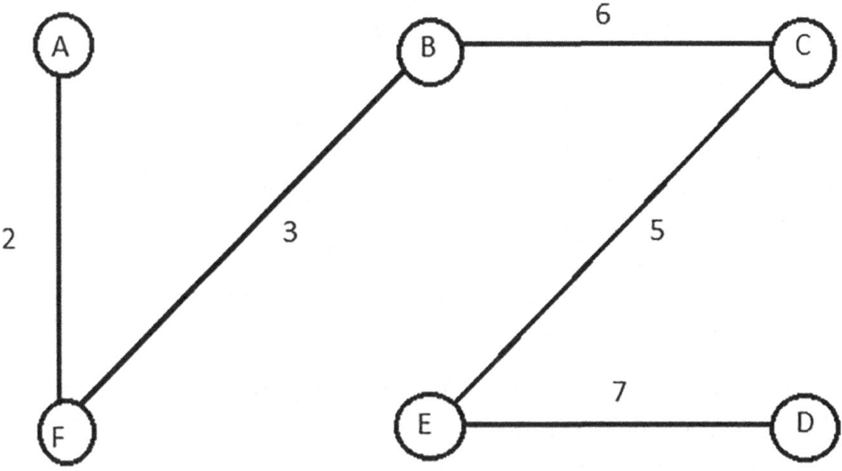

FIGURE 6.38
Minimum cost-spanning tree for graph G in Figure 6.36.

4. In the while loop, the loop repeats itself for the entire edge list. Hence, the complexity would be $O(|E|)$.
5. Complexity of the algorithm would hence be $O(|E| \log_2 |E|)$.

6.5 Interview Questions

1. What is a graph?
2. What are the components that a graph consists of?
3. What is the difference between a connected graph and a non-connected graph?
4. What is the difference between a directed graph and a non-directed graph?
5. What are weighted graphs?
6. How do you represent components of a graph in a computer program?
7. What are the applications of graph data structure?
8. How does depth first traversal work?
9. How does breadth first traversal work?
10. What are the different applications of DFS and BFS?
11. Describe in brief, the terms related to the graph: in-degree, out-degree, nodes and edges.
12. Write an algorithm for BFS on graph.
13. Explain the time complexity of the BFS algorithm.
14. Write a note on the advantages and disadvantages of the BFS algorithm.
15. Explain the in-degree and out-degree of a node with examples.
16. Define the following terms with respect to the graph:
    i. In-degree of a node
    ii. Directed graph
    iii. Weighted graph
    iv. Predecessor
17. Explain the shortest path algorithm for graph with a suitable example.
18. Write a note on MST?
19. Explain in detail the DFS traversal of a graph.
20. Describe non-recursive DFS algorithm in brief.
21. What are the different applications of DFS?

6.6 Multiple Choice Questions

1. Let us consider an unweighted graph G. Let a breadth-first traverse of G be done from a node r. Let d (r, u) and d (r, v) be the lengths of the shortest paths from r to u and v respectively, in G. of u is visited before v during the breadth-first traversal, which of the following statements is correct?

   A. d(r, u)<d (r, v)

   B. d(r, u)>d(r, v)

   C. d(r, u) <= d (r, v)

   D. None of the above

   Answer: (C)

2 How many undirected graphs which are not necessarily connected can be formed out of a provided set V= {V 1, V 2,...V n} of n vertices?

   A. n(n-l)/2

   B. 2^n

   C. n!

   D. 2^(n(n-1)/2)

   Answer: (D)

3 Which of the following statements is/are TRUE for an undirected graph?

   P: Number of odd degree vertices is even
   Q: The sum of degrees of all vertices is even

   A. P Only

   B. Q Only

   C. Both P and Q

   D. Neither P nor Q

   Answer: (C)

4 Consider an undirected random graph of eight vertices. The probability that there is an edge between a pair of vertices is 1/2. What is the expected number of unordered cycles of length three?

   A. A 1/8

   B. 1

   C. 7

   D. 8

   Answer: (C)

5 Given an undirected graph G with V vertices and E edges, the sum of the degrees of all vertices is

A. E

B. 2E

C. V

D. 2V

Answer: (B)

6 How many undirected graphs (not necessarily connected) can be constructed out of a given set V = {v_1, v_2, ... v_n} of n vertices?

A. n(n-1)/

B. 2^n

C. n!

D. $2^{n(n-1)/2}$

Answer: (D)

7. Let G be a weighted undirected graph and e be an edge with maximum weight in G. Suppose there is a minimum weight spanning tree in G containing the edge e. Which of the following statements is always TRUE?

A. There exists a cutset in G having all edges of maximum weight.

B. There exists a cycle in G having all edges of maximum weight

C. Edge e cannot be contained in a cycle.

D. All edges in G have the same weight

Answer: (A)

8. What is the largest integer m such that every simple connected graph with n vertices and n edges contains at least m different spanning trees?

A. 1

B. 2

C. 3

D. n

Answer: (C)

9. Consider a directed graph with n vertices and m edges such that all edges have the same edge weights. Find the complexity of the best-known algorithm to compute the MST of the graph?

A. O(m+n)

B. O(m logn)

C. O(mn)

D. O(n logm)

Answer: (A)

10. For the undirected, weighted graph given below, which of the following sequences of edges represents a correct execution of Prim's algorithm to construct an MST?

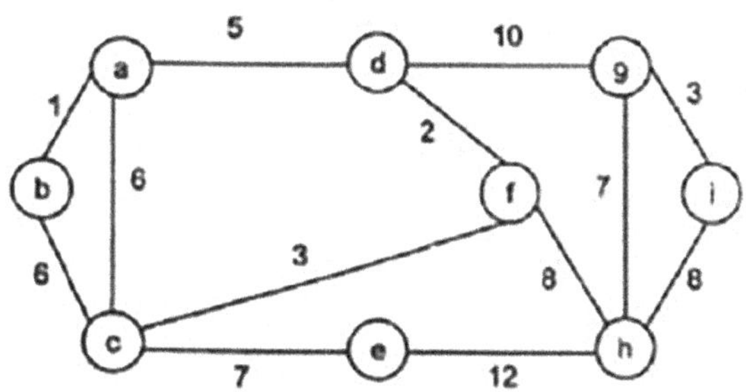

A. (a, b), (d, f), (f, c), (g, i), (d, a), (g, h), (c, e), (f, h)

B. (c, e), (c, f), (f, d), (d, a), (a, b), (g, h), (h, f), (g, i)

C. (d, f), (f, c), (d, a), (a, b), (c, e), (f, h), (g, h), (g, i)

D. (h, g), (g, i), (h, f), (f, c), (f, d), (d, a), (a, b), (c, e)

Answer: (C)

11. What is the number of edges present in a complete graph having n vertices?

A. (n*(n+1))/2

B. (n*(n-1))/2

C. n

D. Information given is insufficient

Answer: B

12. In the given graph, identify the cut vertices.

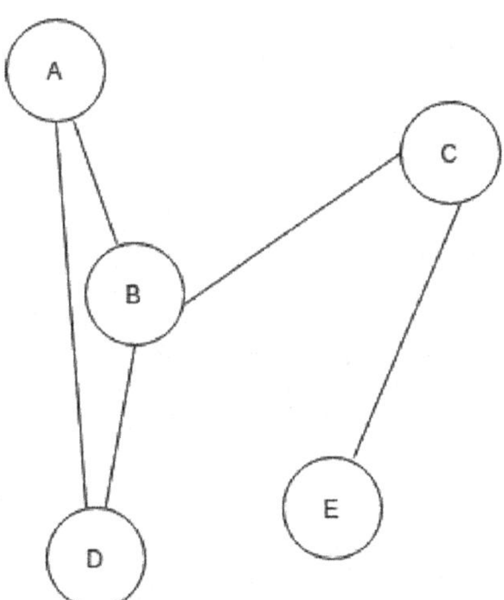

A. B and E

B. C and D

C. A and E

D. C and B

Answer: D

13. In a simple graph, the number of edges is equal to twice the sum of the degrees of the vertices.

A. True

B. False

Answer: B

14. What is the maximum number of edges in a bipartite graph having 10 vertices?

A. 24

B. 21

C. 25

D. 16

Answer: C

15. If a simple graph G, contains n vertices and m edges, the number of edges in the Graph G'(Complement of G) is ___________

A. (n*n-n-2*m)/2

B. (n*n+n+2*m)/2

C. (n*n-n-2*m)/2

D. (n*n-n+2*m)/2

Answer: A

16. A graph with all vertices having an equal degree is known as a ___________

A. Multigraph

B. Regular graph

C. Simple graph

D. Complete graph

Answer: B

17. Which of the following ways can be used to represent a graph?

A. Adjacency list and adjacency matrix

B. Incidence matrix

C. **Adjacency list, adjacency matrix as well as incidence matrix**
D. No way to represent

Answer: C

18. Which of the following is true?
    A. A graph may contain no edges and many vertices
    B. A graph may contain many edges and no vertices
    C. A graph may contain no edges and no vertices
    D. A graph may contain no vertices and many edges

    Answer: B

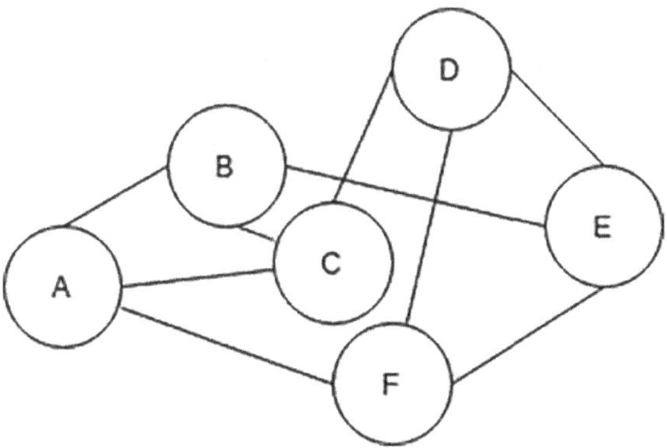

19. The given Graph is regular.
    A. True
    B. False

    Answer: A

20. A connected planar graph having 6 vertices, 7 edges contain ______________ regions.
    A. 15
    B. 3
    C. 1
    D. 11

    Answer: B

21. Which of the following statements is false or true?
    1. If an undirected simple graph of 'n' vertices consists of n (n-1)/2 number of edges then it is called a complete graph.

2. A simple digraph is said to be unilaterally connected if for any pair of nodes of a graph at least one of the nodes of a pair is reachable from the other node.
   A. Statement 1 is false
   B. Statement 2 is false
   C. Statements 1 and 2 are false
   D. Statements 1 and 2 are true

Answer: D

7

Searching and Sorting Techniques

7.1 Need of Searching and Sorting

7.1.1 Need of Searching

Information searching is one of the most important applications of computer science. We often need to find one specific record of data among many hundreds, thousands or more. For example, you might need to find someone's mobile number on your mobile phone or to find the address that requires searching. A particular record can be identified when the key value of that record matches the given input value. If the desired record is found, then the search operation is said to be successful, otherwise searching is unsuccessful. Searching is one of the computer science algorithms. We experience that today's modern computers store a lot of information. To read this information proficiently, we need very efficient searching algorithms. There are certain ways of organizing data that improve the searching process, which means if we keep the information in the right order, it is easy to search the required elements. Sorting is one of the techniques for making the element order.

Definition: Searching refers to the operation of finding the position of a given item in a collection of items. Searching in data structure refers to the process of finding the location of an element in a list.

Two types of searching:

a. **Internal search**

The searching method in which all elements remain sustained in main memory is called internal search.

The time required is less but is perfect for a small amount of data. Linear search, binary search, binary search tree and AVL tree are internal searching techniques.

b. **External search:**

The searching method in which elements are kept in secondary storage is called external search. The time required is longer, and the perfect search technique for a lot of data. B tree and B+ tree are external searching techniques.

DOI: 10.1201/9781003105800-7

7.1.2 Need for Sorting

Sorting is an important step to speed up the subsequent operations on a data structure, like comparing two lists, assigning processes to the processor based on priority, etc. It is easier and faster to search the position of elements in a sorted list than unsorted. Sorting algorithms can be used in a program to sort an array for later searching or writing out of an ordered file or report. Computers spend more time in sorting than any other operation, historically 25% on mainframes. Sorting is the best-studied problem in computer science, with a variety of different algorithms known. Sorting is important because once a set of elements is sorted, many more other problems become easy to solve.

7.2 Sequential Search or Linear Search

7.2.1 Linear Search or Sequential Search Basics

In linear search, we access each element of an array one by one sequentially and see whether or not it is desired element. A search will be unsuccessful if all the elements are accessed, and the desired element is not found.

7.2.2 Efficiency of Linear Search

i. **Best-case time complexity:**

   Searched element is available at the first or almost first place that is at a [0].

   T(n) = Ω (1)

ii. **Worst-case time complexity:**

   Searched element is available at the end of array that is at a [n – 1] or it is not available in the array.

   T(n) = O(n)

iii. **Average-case time complexity:**

   Searched element is present nearly at the middle of the list is called as average case in linear search.

   Average case = (Best case + Worst case)/2

   = (1+n) /2

   T(n) = Θ (n)

Time complexity:

Best-case time complexity	**T(n) = Ω (1)**
Average-case time complexity	**T(n) = Θ (n)**
Worst-case time complexity	**T(n) = O(n)**

7.2.3 Algorithm of Linear Search

1. Start.
2. Traverse the array using any other loop.
3. In every iteration, compare the key element value with the current value of the array.
   i. If the value gets a match, then print the current position or location of the array element.
   ii. If the value does not match, move on to the next array element by incrementing the array index by one.
   iii. If the last element in the array is checked with key elements and no match is found, then print the key element that is not present in your array or list.
4. Stop.

7.2.4 Program of Linear Search

```
#include<stdio.h>
int main()
{
int a[50], n, i, item, loc=-1;
printf("\n Enter size of array or list: ");
scanf("%d",&n);
printf(" Enter values in the array: \n");
         for(i=0;i<n;i++)
         {
         scanf("%d",&a[i]);
         }
printf("\n Enter the number which you want to search: ");
scanf("%d",&item);
         for(i=0;i<n;i++)
         {
         if(item==a[i])
         {
         loc=i;
         break;
         }
         }
         if(loc>=0)
         {
         printf("\n %d is found at the array index %d and at
         position %d",item,loc,loc+1);
         }
         else
         {
         printf("\n Item does not exist in the list");
         }
```

```
return 0;
}
```

Output:

```
 Enter size of array or list: 5
 Enter values in the array:
21
31
44
55
66

 Enter the number which you want to search: 44

 44 is found at the array index 2 and at position 3
```

7.2.5 Advantages of Linear Search

i. It is a very simple method.
ii. It does not require the data to be ordered form.
iii. It does not require any additional data structure.

7.2.6 Disadvantages of Linear Search

i. If several elements are very large, this method of searching is very insufficient and slow.

7.3 Binary Search

7.3.1 Binary Search Basics

Binary search is a very efficient search technique, which works for sorted lists only either in ascending or in descending order. Binary search is a fast searching algorithm with the worst-case time complexity of O ($\log_2$ n). This search algorithm is based on the divide-and-conquer principle. For this algorithm to work properly, the data collection or list must be in sorted order.

7.3.2 Binary Search Algorithm

The logic behind the binary search technique or algorithm is given below:

1. Start.
2. The array must be in ascending or descending sorted order. Then find the middle element of the array, which will be referred to as mid.

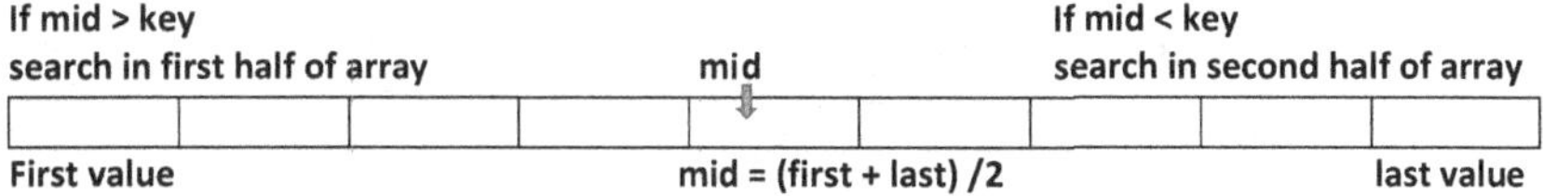

FIGURE 7.1
Binary search.

3. Compare the middle element, mid with searched element or key element, key.
4. There are three cases assumed that the array is in ascending order:
   i. If mid = key, that is, if mid is the desired element, then the search is successful.
   ii. If mid > key, that is, the mid element is greater than key element, then search only in the first half of the array.
   iii. If mid < key, that is, the mid element is less than key element, then search only in the second half of the array, considered array elements are sorted in ascending order.
5. Repeat the above steps 1–3 until the key element is found.
6. Stop.

In this algorithm, every time we reduce the search area. Therefore, the number of comparisons keeps on decreasing. Even in the **worst case**, the number of comparisons is **at most** $\mathbf{log_2 (N+1)}$ (Figure 7.1).

7.3.3 Example of Binary Search

9	12	24	30	45	a
0	1	2	3	4	

Steps to search 45 using binary search in array a [5] are

Step 1: `beg = 0, last = 4`

```
mid = (beg + last)/2 = 2
```

beg		mid		last	
9	12	24	30	45	a
0	1	2	3	4	

a[mid] that is a[2] is 24,
24 < 45 then search in second half of an array, and calculate beg as,
beg = mid+1 = 2+1 = 3 and last=4

Step 2: `mid = (beg + last)/2`

```
= (3+4)/2 = 3
```

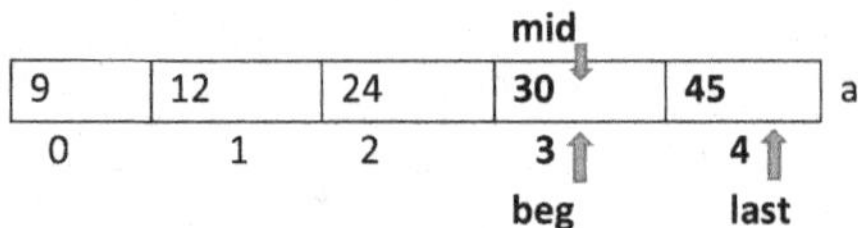

a[mid]<45

Then calculate the beg index as, beg=mid+1=3+1=4 and last=4.

Step 3: `mid = (beg + last)/2`

```
= (4+4)/2 = 4
```

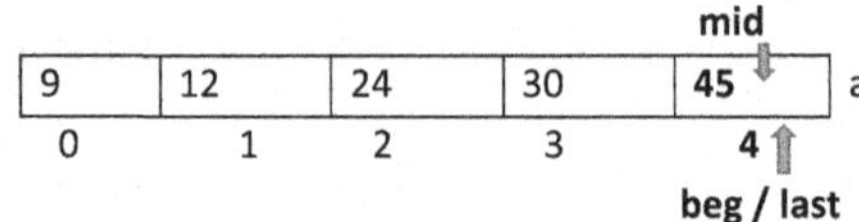

a[mid] that is a[4] is 45,

45=45 then the search is successful at array index number 4 or position fifth of an array.

7.3.4 Program of Binary Search

```
#include<stdio.h>
int main()
{
int a[50] , i , n , loc , mid , beg , end , item , flag=0;
printf("\n Enter the size of array or list:");
scanf("%d",&n);
printf("Enter the elements in array:\n");
        for(i=0;i<n;i++)
        {
        scanf("%d",&a[i]);
        }
        printf("Enter the element which you want to search:");
        scanf("%d",&item);
        loc=0;
        beg=0;
        end=n-1;
while((beg<=end)&&(item!=a[mid]))
{
        mid=(beg+end)/2;
        if(item==a[mid])
        {
        printf("\n Searched element is found");
        loc=mid;
        printf("\n Location of the element is %d and array
        index is %d",loc+1,loc);
```

```
        flag=1;
        }
if(item<a[mid])
        end=mid-1;
else
        beg=mid+1;
}
if(flag==0)
        printf("\n Item is not found in the list.");
return 0;
}
```

Output:

```
 Enter the size of array or list:5
Enter the elements in array:
21
55
77
88
91
Enter the element which you want to search:88

 Searched element is found
 Location of the element is 4 and array index is 3
```

7.3.5 Analysis of Binary Search or Time Complexity of Binary Search

i. **Best case time complexity:**

Searched element is available at the middle of a list that is at a[mid].

T (n) = Ω (1)

ii. **Worst-case time complexity:**

Searched element is available at the end, beginning or not present in the array.

T (n) = O ($\log_2$ n)

iii. **Average-case time complexity:**

Searched element is not available at the end, beginning, or middle of a list, but somewhere else in the array.

T (n) = Θ ($\log_2$ n)

The time complexity of the binary search algorithm is O ($\log_2$ n) as it halves the size of the list at each stage of the search. If a list contains suppose 21 lakhs of elements, then the linear search algorithm requires 21 lakhs key comparisons in the worst-case situation, whereas the binary search algorithm only requires nearly 21 comparisons.

7.3.6 Advantages of Binary Search

1. Worst-case time complexity of the binary search is very efficient.
2. Compared to linear search, the binary search algorithm is much faster.
3. A binary search is a simple algorithm for finding an element in a sorted list.

7.3.7 Disadvantages of Binary Search

1. Does not apply to unsorted records. Data has to be in sorted order always before applying binary search.
2. Binary search is a more complicated algorithm than linear search.
3. Binary search doesn't work well for small lists.
4. Binary search is convenient for array-like direct access structures, but not appropriate for a linked list-like storage structures.

7.4 Analysis of Searching Techniques for Best, Average and Worst Case

There are three different types of searching, in which hashing searching technique gives us the best time complexity rather than linear or binary search technique. The binary search technique gives the best time complexity than that of linear search but elements in binary search must be in sorted order. In all search techniques like linear search, binary search, the time required to search an element depends on the total number of elements in that data structure. In all these search techniques, as the number of elements increases, the time required to search an element also increased linearly.

Hashing is another way of searching, in which time required to search an element does not depend on the number of elements. Using hashing data structure, an element is searched with constant time complexity. Hashing is an effective way to reduce the number of comparisons to search for an element in a data structure. Hashing deals with the idea of proving the direct address of record where the record is likely to store. Hashing is the process of indexing and retrieving key elements or data in a data structure to provide the fastest way of finding the element using the hash key (Table 7.1).

TABLE 7.1

Analysis of Searching Techniques

Analysis of Searching Techniques	Linear Search	Binary Search	Hashing
Best case	Ω (1)	Ω (1)	Ω (1)
Average case	Θ(n)	$\Theta(\log_2 n)$	Θ(1)
Worst case	O(n)	$O(\log_2 n)$	O(1)

7.5 Hashing Techniques

7.5.1 Basic Concept of Hashing and Hash Table

In all searching techniques such as linear search and binary search, the time required to search an element depends on the total number of elements in that data structure. In all these search techniques, the number of elements increases and the time required to search an element also increased linearly.

Hashing is another way of searching, in which time required to search an element does not depend on the number of elements. Using hashing data structure, an element is searched with constant time complexity. Hashing is an effective way to reduce the number of comparisons to search for an element in a data structure. Hashing deals with the idea of proving the direct address of record where the record is likely to store.

Hashing is the process of indexing and retrieving key elements or data in a data structure to provide the fastest way of finding the element using the hash key.

Here, hash key is a value that provides the index value where the actual data is likely to store in the data structure. In this data structure, we use a concept called hash table to store data. All the data values are inserted into the hash table based on the hash key value. The hash key value is used to map the data with the index in the hash table. The hash key is generated for every data using a hash function, which means every entry in the hash table is based on the hash key generated using a hash function.

The hash table is just an array that maps a key element or data into the data structure using a hash function such that insertion, deletion and search operations can be performed with constant time complexity which is equal to O (1).

Hash tables are used to perform the operations such as insertion, deletion and search very quickly in a data structure. Using hash table concept insertion, deletion and search operations are accomplished in constant time. In general, every hash table makes use of a function, which we'll call the hash function to map the data into the hash table.

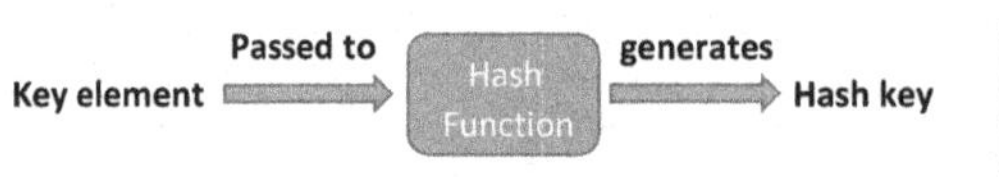

Hash Table

0	Key element or data
1	Key element or data
2	Key element or data
...	Key element or data
n	Key element or data

FIGURE 7.2
Basic concept of hashing and hash table.

A hash function is defined as follows: a hash function takes a piece of data or a key element as input and output an integer, that is, hash key which maps the data to a particular index in the hash table (Figure 7.2).

7.5.2 Example of Hashing

The company has an inventory file that consists of less than 1000 parts. Each part has a unique seven-digit number. The number is called 'key', and a particular keyed record consists of that part name. If there are less than 1000 parts, then a 1000-element array can be used to store the complete file. Such an array will be indexed from 0 to 999. Since the key number is seven digits, it is converted to three digits by taking only the last three digits of a key, which is in Figure 7.3.

If the key is 4,456,000, then that record is stored in the array at the 0th position. That is, the function that converts key, that is seven-digit number, into array position is called **hash function.**

Here, the hash function is,

h(key) = key%1000

key obtained by the hash function is called **hash key** here, 4456000 % 1000 gives us hash key as 0.

7.5.3 Collision in Hash Function

When a hash function returns the same hash address or key for more than one record, this is known as a collision. Note that collision occurrences are a poor design for the hash function.

Position	Key	Record
0	4456000	Data
1	6677001	Data
2	5684002	Data
....	.....	......
333	7676333	Data
334	6677334	Data
335	5678335	Data
....	.....	....
998	1234998	Data
999	2121999	Data

FIGURE 7.3
Hashing.

7.5.4 Rules for Choosing Good Hash Function

1. The hash function should be simple to compute.
2. Number of collisions should be less. Ideally, no collision should occur. Such a function is called **a perfect hash function**.
3. The hash function should produce such keys, which will get distributed uniformly over an array.

   If collisions occur, then they should be handled by applying some techniques, such a technique is called **collision handling or avoidance or resolution technique.**

7.6 Types of Hash Functions

Types of hash functions or hash function calculating methods are as follows:

7.6.1 Mid-Square Method

A good hash function designed with integer key values is the mid-square method. The mid-square method squares the key value and then takes out the middle n bits of the result, giving a value in the range 0 to 2n-1. This works well because most or all bits of the key value contribute to the result.

For example, consider records whose keys are four-digit numbers in base 10. The goal is to hash these key values in a table of size 100, i.e., a range of 0–99. This range is equivalent to two digits in base 10, that is, n=2. If the input is the number 4567, squaring yields an eight-digit number, 20857489. The middle two digits of this result are 57. Now the result value 57 is called as hash key.

Example:

```
Key: 325
325 * 325 = 105625
Hash key: 56
```

7.6.2 Division or Modulus Method

Perhaps the simplest of all the methods of hashing, an integer value key is to divide by n and then to use the remainder as hash key. This is called the division method of hashing or modulus method. In this case, the hash function is as follows:

h(key) = key % m

7.6.3 The Folding Method

The key is divided into different parts, each having the same length as the required address. Then, the pieces are added together, ignoring the carry.

Example:

Key: 3455677234
Partitioning: 345 |567 | 723 |4
Adding: 345+567+723+4=1639

Hash key or address: 639 the carry 1 is ignored.

There are two types of folding methods: one is **fold shift** and another is **fold boundary**. In fold shift, the key value is divided into parts whose size matches the size of the required address or hash key digits. Then the left and right parts are shifted and added with the middle part. In fold boundary, the left and right numbers are folded on a fixed boundary between them and the center number.

i. **Fold shift:**

Key value: 123456789

123+456+789=1368, here 1 is carry and discard or remove the carry digit from obtained address or hash key. Hence, 368 is new hash key or address.

ii. **Fold boundary:**

Key: 123456789

321 (digit reversed)+456+987 (digit reversed)=1764, here 1 is carry and discard or remove the carry digit from obtained address or hash key. Hence, 764 is a new hash key or address.

7.6.4 Digit Analysis

The addresses or hash key is formed by selecting and shifting bits or digits of the original key. The analysis consists of computing the keys and then counting how many times each digit appears in each position. We select those positions where the digits have more uniform distribution. For example, we may select the first, fourth and fifth positions from the original key to form the address or hash key. The function is distribution-dependent. For a given key set, the same rearrangement pattern must be used consistently.

Example:

Keys: 2234, 3452 and 2784

Distribution:

Digit	Position			
	1	**2**	**3**	**4 (from right to left)**
2	1	0	1	2
3	0	1	0	1
4	2	0	1	0
5	0	1	0	0

7	0	0	1	0
8	0	1	0	0

Positions 2 and 3 are best distributed. Hence, we have addressed the hash key as 23, 45 and 78.

7.7 Collision Resolution Techniques

7.7.1 Linear Probing

When a collision occurs that is when two records demand the same location in the hash table, then a collision can be solved by placing the second record linearly down wherever the empty location is found.

Example:

In the hash table given in Figure 7.4, the hash function used is data % 10.

If first element, 131 is placed, then 131% 10=1, that is, 131 placed at index 1 in an array.

If second element 21 is placed, then 21%10=1, that is, 131 is already placed at index 1 which means collision occurs, so we will now apply linear probing. In this method, we will search place for number 21 from the location of 131. In this case, we can place 21 at index 2 and so on.

Because of this technique, searching becomes efficient as we have to search only a limited list to obtain the desired number.

7.7.2 Chaining without Replacement

As a previous method, that is, linear probing, has a drawback of losing the meaning of a hash function, that is, hashing gives us a direct index of the record, the method known as chaining is introduced to overcome this drawback.

Index	data
0	
1	131
2	21
3	31
4	44
5	515
6	61
7	77

FIGURE 7.4
Linear probing.

Index	data	chain
0	-1	-1
1	131	2
2	21	3
3	31	6
4	4	-1
5	5	-1
6	61	-1
7	2	8
8	22	-1

FIGURE 7.5
Chaining without replacement.

Example:
Suppose we have the following elements (Figure 7.5):
131, 21, 31, 4, 5, 61, 2, 22
Explanation:
Here in chaining without replacement method, first 131 element's hash key is generated by using hash function 131% 10, that is, 1. Then 131 is placed in the array index of 1. Then the next key element is 21, and its hash key is generated which is also 1, then place that 21 key element after 131, means 21 is placed at index 2, but in the chain column place, the value of 2 so the key element's having the same hash key is chained together. Similarly, for key element 31 having hash key 1, then place 31 at the index 3 and update the chain value of 21 elements as 3 and so on for next key elements.

7.7.3 Chaining with Replacement

As a previous method, that is, chaining without replacement has a drawback that when the next key element is 2, then the hash function is applied on key element 2, gives us hash key as 2. Then index 2 has already filled but not by the correct key element which is 21. However, we know that 21 is not that position at which currently it is placed. Hence, we will replace 21 by 2, and accordingly, the chain table will be updated. Here the value –1 in the hash table and chain column indicate an empty location (Figure 7.6).

Index	data	chain
0	-1	-1
1	131	7
2	2	8
3	31	6
4	4	-1
5	5	-1
6	61	-1
7	21	3
8	22	-1

FIGURE 7.6
Chaining with replacement.

Advantages of chaining with replacement:

1. The meaning of the hash function is preserved.
2. However, each time some logic is needed to test element, whether it is at its proper position.

7.8 Open and Closed Hashing

7.8.1 Closed Hashing or Open Addressing

Closed hashing is also called open addressing. The first collision resolution method is called closed hashing or open addressing. This method resolves collisions in the home area. When a collision occurs, the home area addresses are searched for key elements for which collision occurs, and that key element is placed at another index in the hash table. In closed hashing, all keys are stored in the hash table itself without the use of linked lists or chains. Open addressing refers to the fact that the location or address of the element is not determined by its hash value exactly and for all time.

Open addressing tells us the index at which an object will be stored in the hash table is not completely determined by its hash key. Instead, the index may vary depending on what is already in the hash table. The "closed" in "closed hashing" refers to the fact that we never leave the hash table; every object is stored directly at an index in the hash table's internal array no separate list is maintained for different hash keys. Note that this is only possible by using some sort of open addressing strategy. Therefore, closed hashing is called open addressing.

Examples of closed hashing or open addressing methods are as follows:

1. Linear probing
2. Quadratic probing
3. Double hashing

1. **Linear probing:**

   In linear probing, we linearly probe for the next slot. When a collision occurs, that is, when two records demand the same location in the hash table, then collision can be solved by placing the second record linearly down wherever the empty location is found.

2. **Quadratic probing:**

   If there is a collision at hash key or address h, quadratic probing method probes the table at locations h+1, h+4, h+9, ...etc., that is, h(key)=key % table size gives us hash key, h then if that location is already filled, then hash key is obtained by h+i^2 where i=1, 2, .. so on gives us an iteration number.

3. Double hashing:

Double hashing uses nonlinear probing by computing different probe increments for different keys. Double hashing uses two hash functions. The first function computes the original address or hash key, and if the slot is available, save the record at that index, but the slot is not empty, then calculate hash key second time using the second hash function.

7.8.2 Open Hashing or Closed Addressing

Open hashing is also called closed addressing. Open hashing or closed addressing is also called separate chaining because, for the same hash key, a separate list is maintained. In open hashing, keys are stored in linked lists, chain, separate list or bucket attached to cells of a hash table. The idea is to make each cell of hash table point to a linked list of records that have the same hash function value.

Examples of open hashing or closed addressing methods are as follows:

1. Chaining
2. Use of buckets

1. Chaining:

One way of resolving collisions is to maintain n linked lists, one for each possible address in the hash table. A key k, hashes to an address i=h (k) in the table. At the address i, we find the head of a list containing all records having keys that have hashed to i. This list is then searched for a record containing key k.

2. Use of buckets:

Suppose we divide a table into n number of groups of records, with each group containing exactly m number of records, then each group of m number of records is called a bucket. The hash function h (k) computes a bucket number from the element key K, and the record containing K is stored in the bucket whose bucket number is h (K). If a particular bucket overflows, an overflow policy is involved.

If a bucket overflows, a chaining technique can be used to link to an "overflow" bucket. This link can be planted at the end of the overflowed bucket. It is convenient to keep overflow buckets on the same cylinder or we may have a separate cylinder for overflows.

7.9 Sorting

Sorting is a basic operation in computer science. Sorting refers to an operation of arranging data in any given sequence, that is, in increasing order or decreasing order.

Internal sorting refers to the process of arranging elements in an array only when they are present in the computer's primary or main memory. External sorting, on the other hand, is the process of sorting elements from an external file by reading them from secondary memory. In internal sorting, all data to be sorted is stored in main memory at all times while sorting is in process. In external sorting, all data to be sorted is stored outside the main memory and only loaded into memory in small chunks as per requirement. External sorting is usually applied in cases when data can't fit into the main memory entirely. The internal sort algorithm has better performance than the external sort.

7.9.1 Bubble Sort

7.9.1.1 Introduction to Bubble Sort

In bubble sort, we have sorted elements in ascending order, and each element is matched with its adjacent element. If the first element is greater than the second one, then the position of elements is interchanged, otherwise, it is not changed. Then the subsequent element is compared with its next adjacent element, and the same process is repeated for all elements in the array.

In the first pass, first, the largest number in the array is placed in the last position in the array.

In the second pass, the second largest number in the array is placed at the second last position in the array and so on.

7.9.1.2 Algorithm of Bubble Sort

1. Start.
2. Read an entire number of elements from user says, n.
3. Stores all elements in the array, which is entered by the user.
4. Specify i=0 for the first pass.
5. Compare the adjacent elements, if (a [j] > a [j+1]) then swap the elements.
6. Repeat step 4 for all the elements.
7. The increment value of i by 1 and repeat steps 4 and 5 till the value of i<n-1.
8. Print the ascending order sorted list of elements.
9. Stop.

7.9.1.3 Example of Bubble Sort

Initial elements in the array without sorting:

11 21 7 35 6

1. First pass:

11 ⤵ No swapping	11	11	11
21	7 ⤵ swapped	7	7
7	21	21 ⤵ No swapping	21
35	35	35	6 ⤵ swapped
6	6	6	**35** largest element in the array placed at last position in array.

2. Second pass:

7 ⤵ swapped	7	7
11	11 ⤵ No swapping	11
21	21	6 ⤵ swapped
6	6	**21** Second largest element in the array placed at second last position in array.
35	**35**	**35**

3. Third pass:

7 ⤵ No swapping	7
11	6 ⤵
6	**11** Third largest element in the array placed at the third last position in array.
21	**21**
35	**35**

4. Fourth pass or last pass:

6 ⤵ swapped. Now remaining six elements are also placed at their sorted position.
7 Fourth largest element in the array placed at fourth last position in an array.
11
21
35

All five-array elements are sorted on the fourth pass. If there are **n** number of elements, then the total number of passes required is **(n-1)**.

7.9.1.4 Program of Bubble Sort

```
#include<stdio.h>
void bubble_sort(int a[],int);
//main function definition
int main()
{
int a[15], num , i;
printf("Enter size of array or list: ");
scanf("%d",&num);
printf("Enter values in the array:\n");
for(i=0;i<num;i++)
{
         scanf("%d",&a[i]);
}
printf("Array elements before sorting:\n");
for(i=0;i<num;i++)
{
         printf("%d\t",a[i]);
}
bubble_sort(a,num);
getch();
}
void bubble_sort(int a[],int n)
{
int i,j,temp;
         for(i=0;i<n-1;i++)
         {
         for(j=0;j<(n-1-i);j++)
         {
         if(a[j]>a[j+1])
         {
         temp=a[j];
         a[j]=a[j+1];
         a[j+1]=temp;
         }
         }
}
printf("\nAfter Bubble sorting array elements are: \n");
for(i=0;i<n;i++)
{
         printf("%d\t",a[i]);
}
return 0;
}
```

Output:

```
Enter size of array or list: 5
Enter values in the array:
21
31
17
89
55
Array elements before sorting:
21      31      17      89      55
After Bubble sorting array elements are:
17      21      31      55      89
```

7.9.1.5 Analysis of Bubble Sort

Consider n is a number of elements in an array. Then, in bubble sort, first pass requires (n-1) comparisons to fix the highest element to its proper location. Second pass requires (n-2) comparisons and so on.

Therefore, the total comparisons required are as follows:

```
f(n) = (n-1) + (n-2) + (n-3) + ..............3+2+1 = n(n-1)/2
f(n) = O(n²)
T(n) = O(n²)
```

Thus, the time complexity of the bubble sort algorithm is **O (n^2)**. For bubble sort algorithm, worst case, average-case and best-case time complexity remains the same, that is, n^2.

Space complexity of bubble sort:

O (1) is the space complexity of bubble sort.

7.9.1.6 Advantages of Bubble Sort

1. Bubble sort is a simple sorting method.
2. No additional data structure is necessary.
3. Items are exchanged in place without the use of additional temporary storage, so the required space is minimal.

7.9.1.7 Disadvantages of Bubble Sort

1. Bubble sort is a very inefficient method of sorting.
2. Even if the elements are in sorted order, all (n-1) passes will be done.
3. The bubble sort is mostly suitable for academic teaching but not for real-life applications.
4. Bubble sort does not deal well with a list containing a huge number of elements.

7.9.2 Selection Sort

7.9.2.1 Introduction to Selection Sort

Selection sort is also known as **a push-down sort.** Here, we have sorted elements in ascending order. The selection sort consists entirely of a selection phase in which the smallest element, in the array is searched. Once the smallest element is found, it is placed in the first position of the array, that is, at index 0. The other remaining elements are made to find the next smallest element, which is placed in the second position of the array, that is, at index 1 of the array and so on.

Thus, all elements are sorted into ascending order.

Let a be an array of n elements.

Pass 1: Find the position i of the smallest element in the list of n elements a[0], a [1],………a [n-1] and then interchange a[i] with a[0], if index 0 and i are different otherwise no swap. Thus, a[0] is sorted.

Pass 2: Then find the second smallest element in the list of (n-1) elements from a [1], a [2],………, a [n-1] that is a [i] then interchange a [i] with a [1] if index 1 and i are different otherwise no swap. Then a [0] and a [1] are sorted.

Pass 3: Find the third smallest element in the list of (n-2) elements from a [2], a [3],……..,a [n-1] that is supposed a [i] then interchange a[i] with a[2] if index 2 and i are different otherwise no swap. Then a [0], a [1], a [2] are sorted and so on up to pass (n-1).

Thus, array a is sorted after (n-1) pass.

7.9.2.2 Example of Selection Sort

Initial array a contains four elements: 21, 5, 9, 2. a[3]

	21	5	9	2	a
index	0	1	2	3	

Here, in the first pass, find the smallest element, that is 2, at index 3. In addition, 0 index and index 3 are different; then, swap the values of these two indices that is a [0] which is 21 and a [3] which is 2. Now element 2 is sorted to its own position. Therefore, after completion of first pass array, a becomes as follows:

	2	5	9	21	a
index	0	1	2	3	

In the second pass, find the smallest element in the array a from index 1 to 3. Here 5 is the smallest element in the array a from index 1 to 3. Therefore, check the index of minimum element and 1, here both are same, so no swap requires. Now elements 2 and 5 are sorted to their own position. Therefore, after completion of the second pass array, a becomes as follows:

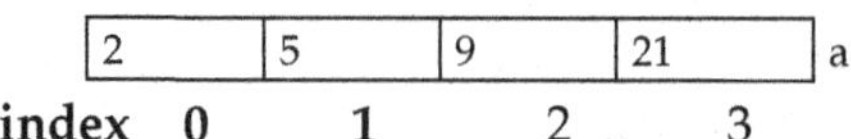

In a third pass, find the smallest element in the array a from index 2 to 3. Here 9 is the smallest element in the array a from index 2 to 3. Therefore, check the index of the minimum element and 2, here both are same, so no swap requires. Now elements 2, 5 and 9 are sorted to their own position. Therefore, after completion of the third pass array, a becomes as follows in which all elements are in the sorted position. Here total array elements are four, so total three passes are required to sort the array:

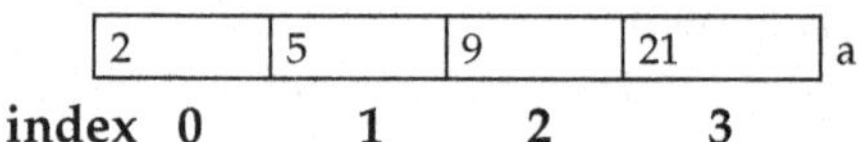

7.9.2.3 Program of Selection Sort

```
#include<stdio.h>
void selection_sort(int *, int);
int main()
{
int a[10], num , i;
printf("\nEnter size of array or list: ");
scanf("%d",&num);
printf("Enter values in the array: \n");
for(i=0;i<num;i++)
{
        scanf("%d",&a[i]);
}
printf("Array elements before sorting: \n");
for(i=0;i<num;i++)
{
        printf("%d\t",a[i]);
}
selection_sort(a, num);
return 0;
}
void selection_sort(int *a,int n)
{
int i,j,loc,min,temp;
for(i=0;i<n-1;i++)
{
        min=a[i];
        loc=i;
        for(j=i+1;j<n;j++)
        {
                if(min>a[j])
                {
                min=a[j];
```

```
                    loc=j;
                    }
        }
if(loc!=i)
{
temp=a[i];
a[i]=a[loc];
a[loc]=temp;
}
}
printf("\nArray elements after selection sort:\n");
for(i=0;i<n;i++)
{
printf("%d\t",a[i]);
}
}
```

Output:

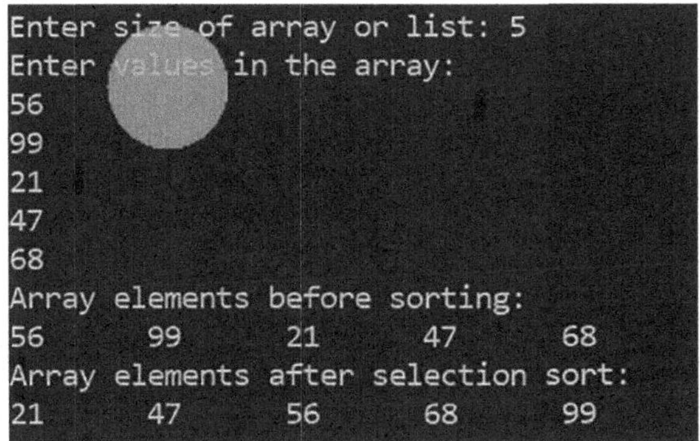
Enter size of array or list: 5
Enter values in the array:
56
99
21
47
68
Array elements before sorting:
56 99 21 47 68
Array elements after selection sort:
21 47 56 68 99

7.9.2.4 Analysis of Selection Sort

In selection sort, first pass requires (n-1) comparisons to fix the highest element to its proper location. Second pass requires (n-2) comparisons and so on.

Therefore, total comparisons are,

$$f(n) = (n-1) + (n-2) + (n-3) + \ldots\ldots 3+2+1 = n(n-1)/2$$
$$f(n) = O(n^2)$$
$$T(n) = O(n^2)$$

Thus, the time complexity of the selection sort algorithm is $O(n^2)$. For the selection sort algorithm, worst-case, average-case and best-case time complexity remains the same, that is, n^2.

Space complexity of selection sort:

O (1) is the space complexity of the selection sort.

7.9.2.5 Advantages of Selection Sort

1. Selection sort is a simple sorting method.
2. No additional data structure is required.
3. Selection sort is an in-place sorting algorithm, and no additional temporary storage is required to sort the elements in that particular list.

7.9.2.6 Disadvantages of Selection Sort

1. Selection sort is a very inefficient method of sorting.
2. Even if the elements are in sorted order, all (n-1) passes will be done.
3. Selection sort is poor efficiency when dealing with a huge list of elements.

7.9.3 Insertion Sort

7.9.3.1 Introduction to Insertion Sort

An insertion sort is one that sorts a set of values by inserting values into an existing sorted file. In the insertion sort algorithm, every pass moves an element from unsorted portion to sorted portion until all the elements are sorted in the list.

7.9.3.2 Example of Insertion Sort

Consider an array a[5] contain five elements.

45	-30	170	97	21
a[0]	a[1]	a[2]	a[3]	a[4]

Pass 1: First copy a[1] value into temp variable. Compare temp with a[0], if temp<a[0], then interchange the elements.

-30	45	170	97	21
a[0]	a[1]	a[2]	a[3]	a[4]

Pass 2: Now copy a[2] value into temp variable. Compare temp with a[1] and a[0] if temp>a[1] and a[0], then no need to interchange.

-30	45	170	97	21
a[0]	a[1]	a[2]	a[3]	a[4]

Pass 3: For the next pass, copy a[3] value into temp variable. Then compare temp with a[2], a[2]>temp, then shift the value of 170 to the a[3] array

position. Again compare temp with a[1], here a[1]<temp so insert the temp value into the array position a[2].

-30	45	97	170	21
a[0]	a[1]	a[2]	a[3]	a[4]

Pass 4: For the next pass, copy a [4] value into temp variable. Then compare temp with a [3], a [3]>temp, then shift the value of 170 to the a [4] array position. Again, comparing a [2] with temp, here a [2]>temp so shift the value 97 to the a [3] array position. Again, comparing a [1] with temp, here a [1]>temp so shift the value 45 to the a [2] array position. Again, comparing a [0] with temp, here a [0]<temp so insert the temp value into the array position a [1]. Therefore, in this sort, we have to insert the new value into the already sorted array so-called as insertion sort. After insertion of value 21 into array position a [1], all array elements are sorted in ascending order as shown in the below figure.

-30	21	45	97	170
a[0]	a[1]	a[2]	a[3]	a[4]

Thus, elements of array a are sorted by insertion sort.

7.9.3.3 Program of Insertion Sort

```
#include<stdio.h>
// main function definition
int main()
{
int a[10], n, i, j, k, temp;
printf("Enter size of array or list: \n");
scanf("%d",&n);
printf("Enter values in the array: \n");
for(i=0;i<n;i++)
{
         scanf("%d",&a[i]);
}
printf("Array elements before sorting: \n");
for(i=0; i<n; i++)
{
         printf("%d\t", a[i]);
}
//insertion sort
//n = number of elements in array.
//temp = temporary variable which we want to insert into
already sorted array
// k= total number of passes.
//j = control variable used for internal comparison
```

```
for(k=0; k<n; k++)
{
          temp = a[k];
          j=k-1;
          while((a[j]>temp)&&(j>=0))
          {
          a[j+1]=a[j];
          j--;
          }
a[j+1]=temp;
}
printf("\nArray elements After insertion sort:\n");
for(i=0;i<n;i++)
{
          printf("%d\t", a[i]);
}
return 0;
}
```

Output:

```
Enter size of array or list:
5
Enter values in the array:
21
71
33
97
45
Array elements before sorting:
21        71        33        97        45
Array elements After insertion sort:
21        33        45        71        97
```

7.9.3.4 Analysis of Insertion Sort

i. **Best case time complexity of insertion sort:**

   Best case occurs when the array is already in sorted form. In this case, run time control will not enter into the while loop.

   Then, **T (n) = 1 + 1 + ... + 1 (n times)**

   T (n) = Ω (n)

ii. **Worst case time complexity of insertion sort:**

   The worst case occurs when an array is in reverse order or almost all array elements are in unsorted array positions.

   Then, **T (n) = 1+2+3+............. + (n-1).**

   = n (n-1)/2.

   T (n) = O (n^2)

iii. **Average-case time complexity of insertion sort:**

Average case occurs when half of array elements are unsorted positions.

Then, **T (n) = 1+2+3+............ + (n-1).**

= n (n-1)/2.

T (n) = Θ (n²)

$$T(n) = 1+2+3+\ldots\ldots + (n-1) = n(n-1)/2, \quad T(n) = \Theta(n^2)$$

Space complexity of insertion sort:

O (1) is the space complexity of insertion sort.

7.9.3.5 Insertion Sort: Real-Time Example

Arrangement of card-by-card player. The card player picks up the cards and inserts them into the required position. Thus, at every step, we insert an item in its proper place in an already ordered list.

7.9.3.6 Advantages of Insertion Sort

1. Insertion sort is a simple sorting method.
2. No additional data structure is required.
3. For best case, time complexity is **Ω (n)**.
4. Insertion sort gives a good performance when dealing with a small list.
5. The insertion sort is an in-place sorting algorithm so the space requirement is minimal.

7.9.3.7 Disadvantages of Insertion Sort

1. Insertion sort is an inefficient method when elements are almost unsorted. At that time, complexity is **O (n²)**.
2. As the number of elements increases in the insertion sort, the performance of the algorithm gets affected and becomes slow. Insertion sort needs a large number of element shifts.

7.9.4 Quick Sort

7.9.4.1 Introduction to Quick Sort

Quick sort introduced C.A.R. Hoare in 1962 is a very fast method of internal sorting.

It is based on the divide-and-conquer paradigm.

7.9.4.2 Algorithm of Quick Sort

1. Start.
2. Read the total number of the elements from the user say, n.

3. Stores all the elements in the array, which is entered by the user.
4. Take the first element from the array and called it the pivot element.
5. Now find all elements, which are less than this pivot element and place them before the pivot elements.

Thus, an array will be divided into:

i. Lesser elements
ii. Pivot elcments
iii. Elements greater than the pivot element

6. Repeat steps 3 and 4 each time placing a pivot element at its proper position. This complete list will get sorted.
7. Stop.

7.9.4.3 Example of Quick Sort

An initial array of size 5 is shown as follows:

25	57	48	38	10
a[0]	a[1]	a[2]	a[3]	a[4]

Pass 1: set a[0], the element that is 25 as the pivot element. Now find all the elements lesser than 25 and place 25 after all lesser elements.

10	25	57	48	38
sub-list 1	pivot		sub-list 2	

Pass 2: Now sub-list 1 having only one element then no need to sort it, but sub-list 2 can be sorted. For that, set the pivot element as 57.
Find all lesser than 57 elements and place them before 57.

Pass 3: Again from sub-list 1 48 is a new pivot element. Place all lesser than 48 elements before it.

10	25	38	48	57
		sub-list 1	pivot	

Here, the sub-list 1 contains only one element and lists having one element each so stop the algorithm.
Thus, the list is sorted by quick sort.

7.9.4.4 Program of Quick Sort

```
#include<stdio.h>
void quickSort(int list[], int, int);
//main function definition
int main()
{
int list[20], size, i;
printf("Enter size of the list: ");
scanf("%d", &size);
printf("Enter %d integer values: \n", size);
for(i = 0; i < size; i++)
{
        scanf("%d", &list[i]);
}
printf("Array elements before sorting:\n ");
for(i = 0; i < size; i++)
{
        printf("%d\t", list[i]);
}
quickSort(list, 0, size-1);
printf("\nArray elements after sorting:\n ");
for(i = 0; i < size; i++)
{
        printf("%d\t", list[i]);
}
return 0;
}
void quickSort(int list[],int low, int high)
{
int pivot, l, h, temp;
if(low < high)
{
pivot = low;
l = low;
h = high;
while(l < h)
{
        while(list[l] <= list[pivot] && l < high)    // l <
high condition means increment l till it reach
                                                      //to the
high index of that sub-list
        {
        l++;
        }
        //here low and h is not compared because list[pivot]
        element means list[low] element //and both are same so
        h never becomes less than the low so low < h condition
        does //not checked like previous one.
        while(list[h] > list[pivot])
```

```
        {
        h--;
        }
//swap the list[l] and list[h] element
if(l < h)
{
        temp = list[l];
        list[l] = list[h];
        list[h] = temp;
}
} //completion of outer while loop
//swap the element list[pivot] and list[h]
temp = list[pivot];
list[pivot] = list[h];
list[h] = temp;
quickSort(list,low,h-1);    //sort left sub-list
quickSort(list,h+1,high);   //sort right sub-list
}   // completion of if
}
```

Output:

```
Enter size of the list: 5
Enter 5 integer values:
55
77
99
22
58
Array elements before sorting:
 55      77      99      22      58
Array elements after sorting:
 22      55      58      77      99
```

7.9.4.5 Analysis of Quick Sort

Quick sort time complexity is calculated as follows:

T (n) = {O (n) + T(n/2) + T (n/2) }

O (n) = time required to partition array.

T (n/2) = time required to sort the left sub-array.

T (n/2) = time require to sort the right sub-array.

i. **Worst-case time complexity of quick sort:**

In worst-case time complexity, on every function calls given array is portioned into two sub-arrays. One of them is an empty array, then it is called a worst case.

$$\mathbf{T(n) = O(n) + T(0) + T(n-1)}$$
$$\mathbf{= O(n) + O(n-1) + T(n-2)}$$

$= O(n) + O(n-1) + O(n-2) + T(n-3)$
$= O(n) + O(n-1) + O(n-2) + \ldots + O(1)$
$= O(n(n+1)/2)$
$= O(n2)$

ii. **Best case time complexity of quick sort:**

Best case timing analysis is possible when the array is always partition in two almost equal halves.

$T(n) = \Omega(n) + T(n/2) + T(n/2)$
$= \Omega(n) + 2T(n/2)$
$T(n) = \Omega(n \log 2\, n)$

Note that $\log_2 n$ because $\log_2 (4)$ gives you value is 2 that is each time sub-list is divided into two nearly equal halves in each pass. Therefore, time complexity becomes $T(n) = \Omega(n \log_2 n)$ for best case.

Thus the best case partitioning produces a much faster algorithm.

iii. **Average-case time complexity of quick sort:**

Average-case timing analysis is possible when the array is nearly half time partition in two almost equal halves.

$T(n) = \Theta(n) + T(n/2) + T(n/2)$
$= \Theta(n) + 2T(n/2)$
$T(n) = \Theta(n \log 2\, n)$

Space complexity of quick sort:

$O(\log_2 n)$ is the space complexity of quick sort.

7.9.4.6 Advantages of Quick Sort

1. The performance of the quick sort CPU cache is superior to that of other sorting algorithms. This is because of its in-place characteristic. The CPU cache is a piece of hardware that reduces the access time to the data in the memory by keeping some part of the frequently used data of the main memory in itself. It is smaller and faster than the main memory. Therefore, quick sort is the fastest sorting algorithm.
2. No additional data structure is required like merge sort or heap sort.
3. Quick sort is superior to merge sort in terms of space is concerned. Space or memory required for the quick sort is less than that of the merge sort.
4. Quick sort is in-place sorting algorithm, whereas merge sort is not in-place. In-place sorting means, it does not use additional storage space to perform sorting. In merge sort, to merge the sorted arrays, it requires a temporary array and hence it is not in-place.
5. Quick sort is better than other sorting algorithms with the same time complexity O (n log2 n) that is merge sort and heap sort. Even though quick sort has $O(n^2)$ in the worst case, it can be easily avoided with high probability by selecting the correct pivot element.

7.9.4.7 Disadvantages of Quick Sort

1. Time efficiency of the quick sort depends on the selection of the pivot element. Inadequately picked pivot element leads to the worst-case time complexity.
2. Tough to implement a partitioning algorithm in quick sort.

7.9.5 Merge Sort

7.9.5.1 Introduction to Merge Sort

Merge sort is based upon the divide and conquer algorithm. In merge sort, we divide the main list into two sub-lists, then go on dividing those sub-lists till we get sufficient length of that sub-list. Then we compare and sort the elements of each list. Merge the two sub-lists and sort the merged list. This process will be repeated until we get one sorted list.

7.9.5.2 Algorithm of Merge Sort

Algorithm mergeSort (arr[], low, high)
If high>low

1. Find the central point to partition the array into two halves:
   Middle index of array or list, mid=(low+high)/2
2. Call mergeSort for the left or first half:
   Call mergeSort (arr, low, high)
3. Call mergeSort for the right or second half:
   Call mergeSort (arr, mid+1, high)
4. Merge the two halves sorted in steps 2 and 3:
   Call merge (arr, low, mid, high)

7.9.5.3 Example of Merge Sort

Step number indicates the order in which steps are get executed.

91	-38	48	33	56	25	10	84
a[0]	a[1]	a[2]	a[3]	a[4]	a[5]	a[6]	a[7]

Step 1:
The elements are stored in an array, we will go on dividing the array.

91	-38	48	33
a[0]	a[1]	a[2]	a[3]

56	25	10	84
a[4]	a[5]	a[6]	a[7]

Step 2: step 12

Again divide each sub list.

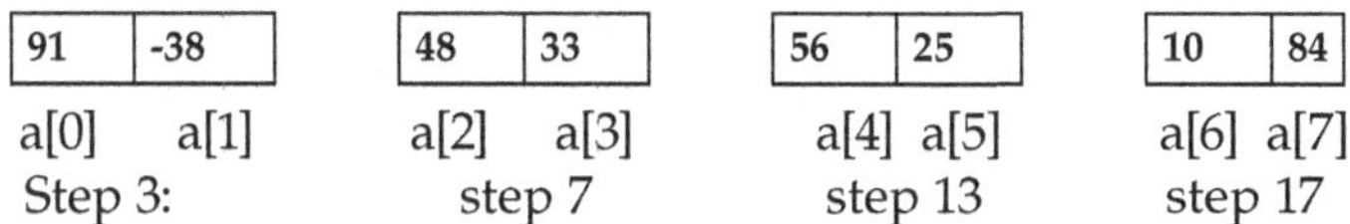

Now if we try to divide the sub-lists, we will get single element in each list.

91	-38	48	33	56	25	10	84
a[0]	a[1]	a[2]	a[3]	a[4]	a[5]	a[6]	a[7]
Step 4	step 5	step 8	step 9	step 14	step 15	step 18	step 19

Further division is not possible. Hence, we will stop dividing and start merging.

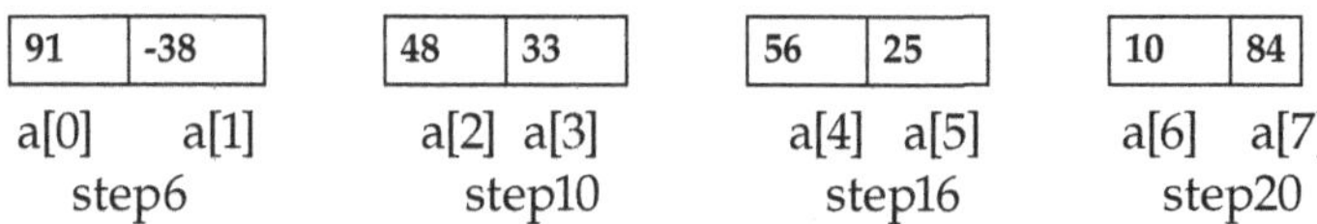

Then sort that sub-list only.

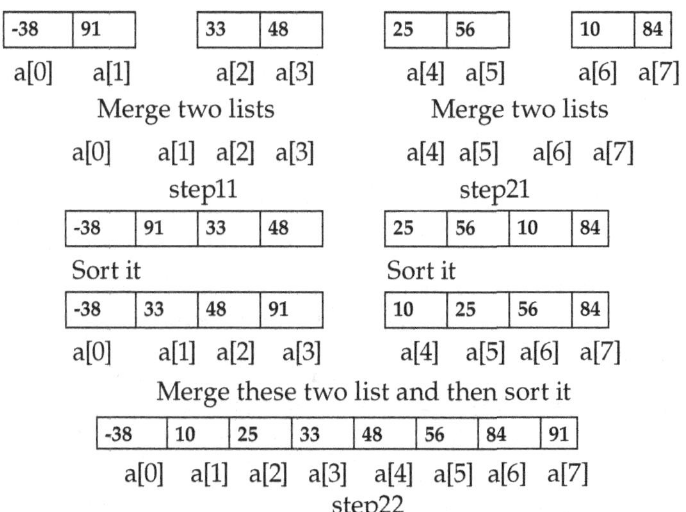

Here is the sorted list.

7.9.5.4 Program of Merge Sort

```
#include<stdio.h>
//function prototypes
void merge_sort(int *a, int *t, int low, int high);
```

```
void merge(int *a, int *t, int low, int mid, int high);
//main function definition
int main()
{
int n, a[10], t[10], l=0, h, i;
printf("Enter size of array:");
scanf("%d", &n);
printf("Enter %d array elements:\n", n);
h=n-1;
for(i=0;i<=(n-1);i++)
{
        scanf("%d", &a[i]);
}
printf("Array elements before sorting:\n");
for(i=0;i<=(n-1);i++)
{
        printf("%d\t", a[i]);
}
merge_sort(a,t,l,h);
printf("\nArray elements after sorting:\n");
for(i=0;i<=(n-1);i++)
{
        printf("%d\t",a[i]);
}
return 0;
}
//function definition for merge sort
void merge_sort(int *a, int *t, int low, int high)
{
int mid;
if(high>low)
{
mid=(high+low)/2;  //find the middle array index
merge_sort(a,t,low,mid); //call to the left sub-list till left
sub-list contains only one element
merge_sort(a,t,mid+1,high); // call to the right sub-list till
right sub-list contains only one element
merge(a,t,low,mid+1,high); //call to merge and sort function
}
}
// function definition for merging two sub-lists that are left
and right sub-lists
void merge(int *a, int *t, int low, int mid, int high)
{
//mid is the first element's index of the right sub-list
//high is the last element's index of the right sub-list
//pos variable is refers index of temporary array t
// a is the original array
int pos, n, end, i;
pos = low; //low is first element's index of the left sub-list
```

```
n = (high-low + 1); // n stores total number of elements
present in left and right sub-list
end = mid-1;          // end is the last element's index of
left sub-list
//copy elements of left and right sub-list into temp array t
while(low<=end && mid<=high)
{
/* copy element of left or right sub-list into temporary array
t depending on the value of the left or right sub-list. If the
value of the first element in the left sub-list is low than
that of the right sub-list the copy the first element of the
left sub-list into the temporary array t. Thus temporary array
contains elements in sorted order. Continue the while loop
till low<=end && mid<=high */
        if(a[low]<a[mid])
        {
        t[pos]=a[low];
        pos++;
        low++;
        }
        else
        {
        t[pos]=a[mid];
        pos++;
        mid++;
        }
}
// If left sub-list contains some elements, then just copy
those remaining elements into temporary //array t
while(low<=end)
{
        t[pos]=a[low];
        pos++;
        low++;
}
// If right sub-list contains some elements, then just copy
that remaining elements into temporary //array t
while(mid<=high)
{
        t[pos]=a[mid];
        pos++;
        mid++;
}
//copy all elements in the temporary array t into the original
array a from high index to the number of //elements present in
that temporary array. n variable gives us the total number of
elements present //in the left and right sub-list
for(i=0 ; i<n ; i++)
{
        a[high]=t[high];
```

```
        high--;
}
}
```

Output:

```
Enter size of array:5
Enter 5 array elements:
79
67
21
99
32
Array elements before sorting:
79      67      21      99      32
Array elements after sorting:
21      32      67      79      99
```

7.9.5.5 Analysis of Merge Sort

This sorting method is quite efficient. It requires ($\log_2 n$) passes and O ($n \log_2 n$) comparisons.

Even in the worst case it never requires more than ($n \log_2 n$) comparisons.

The only drawback in this method is the requirement of O (n) additional space for the temporary array.

Time complexity of merge sort:

1. Best case=Ω (n log n)
2. Average case=Θ (n log n)
3. Worst case=O (n log n)

Space complexity of merge sort:

O (n) is the space complexity of merge sort.

7.9.5.6 Advantages of Merge Sort

1. Even worst-case time complexity of merge sort is ($n \log_2 n$) which does not change.
2. Merge sort can be applied to files of any size.
3. Good for sorting slow-access data, for example, tape drive or hard disk.
4. It is excellent for sorting data that are normally accessed sequentially. For example, linked lists, tape drive, hard disk and receiving online data one item at a time

7.9.5.7 Disadvantages of Merge Sort

1. Additional memory space is required for the temporary storage of divided arrays.
2. Merge sort is less efficient than other sorts.
3. Extra data structure is required to store temporary array.

7.9.5.8 Applications of Merge Sort

1. Merge Sort is useful for sorting the linked list in O (n Log2 n) time.
2. Merge Sort is used in external sorting.
3. Merge Sort is used in e-commerce applications.

7.9.6 Heap Sort

7.9.6.1 Introduction to Heap Sort

Heap sort is a non-linear sort.

Heap sort is based on a heap structure which is a special type of binary tree. Heap sort consists of two phases:

1. **Creation of heap**
2. **Processing of heap**

First, we have to define a heap,

A heap of size 'n' is a binary tree of 'n' nodes that satisfies the following two constraints:

1. The binary tree is an **almost complete binary tree**. That is, each leaf node in the tree is either at level d or at level d-1 that means the difference between levels of leaf nodes is only one is permitted. And at depth d, that is the last level, if a node is present, then all the nodes to the left of that node should also be present first, and no gap between the two nodes from left to the right side of level d.
2. Keys in nodes are arranged such, that content or value of each node is less than or equal to the content of its father or parent node. Which means for each node info [i] <= info [j], where j is the father of node i. This heap is called as max heap.

Here, in heap, each level of binary tree is filled from left to right and a new node is not placed on a new level until the preceding level is full.

7.9.6.2 Examples of Heap

We may note that the root of the heap contains the largest value element. Also, any path from the root to a leaf is an ordered list in descending order (Figure 7.7).

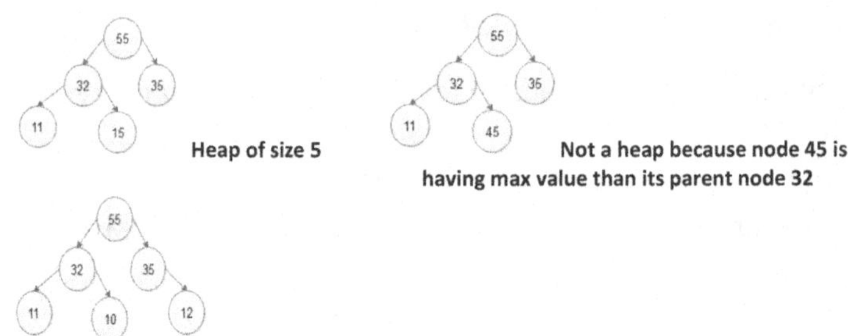

Nota heap because at level 2, 35 node has a right child but no left child

FIGURE 7.7
Examples of a Heap and not a Heap.

7.9.6.3 Creating a Heap

The unsorted keys are taken sequentially one after the other and added into a heap. The size of heap grows with the addition of each key.

The ith key that is, k_i is added into a present heap of size (i−1) and a heap of size i is generated.

Initially, the node is arranged in the heap of size (i−1) so that the almost complete constraint is fulfilled.

The value of the key, ki is then compared to the value of its parent key.

If ki is greater, the contents of newly added node and that a parent node are interchanged. This process keeps on until either ki is at the root node or parent's key value is not less than ki. The final tree is a heap of size i.

Here, the resulting heap is stored in the array level by level, from left to right.

The root is stored in a heap [0] and the last node in the heap [maxnodes-1] where max nodes is the number of nodes in the heap.

We can note that for any node heap [i], its two children are residing in a heap [i*2] and a heap [i*2+1].

If we want to see the parent of any heap [k], we can get it at the heap [k/2].

Now let's write, an algorithm for creating a heap of size i by adding a key to a heap of size (i – 1) where i >=1.

Algorithm of creating a heap:

1. Start
2. `s=i;`
3. Find parent node index of i th node in the array as,
```
parent = s / 2;
key[ s] = newkey ;
```
4. `while ( s != 0 && key[parent] <= key[s] )`
```
{
 Thereafter exchange the parent and the child nodes as,
  temp = key [parent] ;
```

```
         key[parent] = key[s];
         keys[s] = temp;
       Advance the new node one level up in the tree as
follows,
       s = parent;
       parent = s / 2 ;
       }
```

5. Stop.

Example of creating a heap: Consider the following unsorted list of keys:
10, 20, 9, 4, 15, 17

Heap of size 1

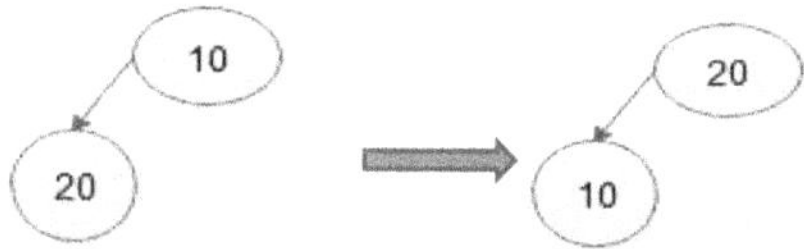

Heap of size 2

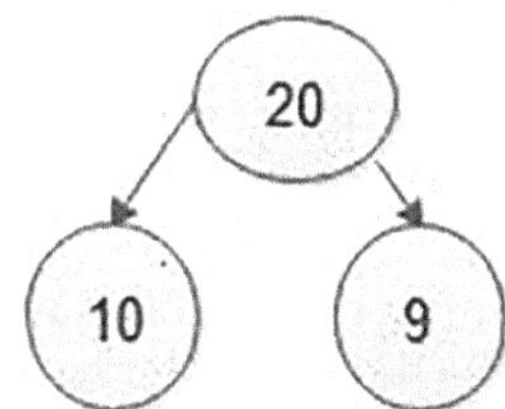

Heap of size 3

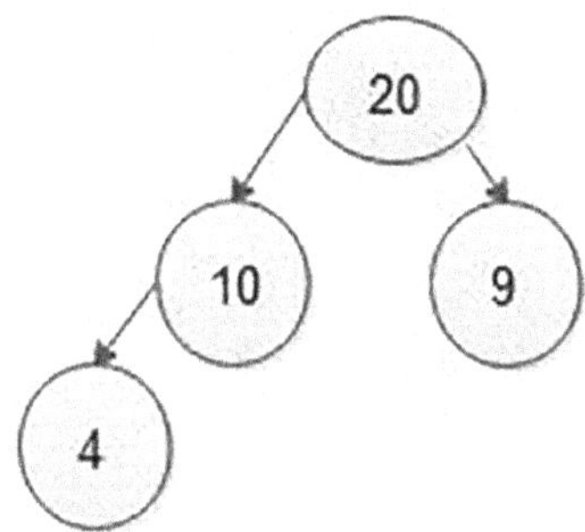

Heap of size 4

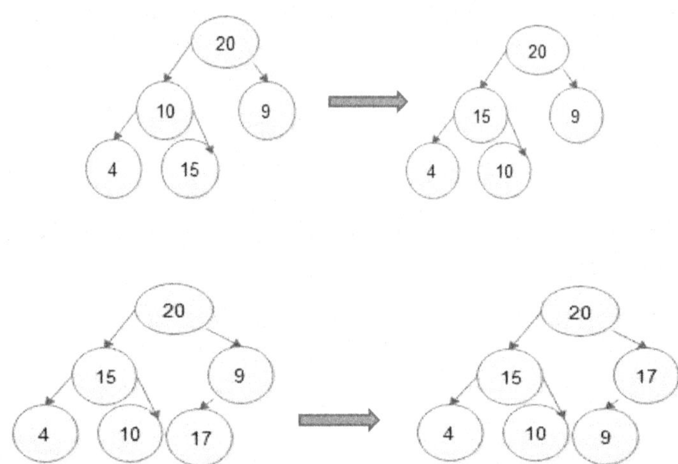

FIGURE 7.8
Creating of a heap of size 6.

Heap of size 5
Heap of size 6
See Figure 7.8.

7.9.6.4 Processing a Heap

This heap may be represented as an array in the below figure (Figure 7.9),

We may note that the resulting heap depends on the initial ordering of the unsorted list.

For different order of the input list, the heap would be different.

Now, we have to process the heap in order to generate a sorted list of keys.

We know that the largest element is at the top of the heap which is sorted in the array at position heap [0]. We interchange heap [0] with the last element in the heap array heap [maxnodes-1] so that the heap [0] is in its proper place. We then adjust the array to be a heap of size (n – 1). Again interchange

Index	value
0	20
1	15
2	17
3	4
4	10
5	9

FIGURE 7.9
Heap of size 6 is represented using an array.

heap [0] with heap [n – 2], adjusting the array to be a heap of size (n – 2) and so on. At the end, we get an array that contains the keys in sorted order. The algorithm to process a heap is as follows:

Algorithm of processing a heap:

Step 1: Start.

Step 2: Interchange the root node with the last node in the heap array.

Step 3: At present, heap [maxnode-1] is in its correct position.

Step 4: Now, compare the new root value with its left child value.

Step 5: If the new root value is smaller than its left child, then compare the left child with its right sibling. else go to Step 7.

Step 6: If the left child value is larger than its right sibling, then swap root with the left child. Otherwise, swap root with its right child.

Step 7: If the root value is larger than its left child, then compare the root value with its right child value.

Step 8: If the root value is smaller than its right child, then swap root with the right child. Otherwise, stop the process.

Step 9: Repeat the same until the root node is fixed at its exact position.

Step 10: Repeat the step 2 to 9 for new root till the heap contains one element.

Step 11: Stop.

Example of processing a heap:

After processing of heap array becomes as follows (Figure 7.10):

7.9.6.5 Program of Heap Sort

```
#include <stdio.h>
#define LENGTH 10
int heapSize, n;
void Heapify (int* A, int i)
{
```

Index	value
0	4
1	9
2	10
3	15
4	17
5	20

FIGURE 7.10
Heap array of size 6 after processing of heap.

```
int l = 2 * i + 1;  //index of left child
int r = 2 * i + 2;  //index of right child
int largest;

if(l <= heapSize && A[l] > A[i]) //compare value of left child
with value of parent node
largest = l;    //if left child's value is greater assign left
child index to the largest variable
else
largest = i;   //Otherwise parent node's value is greater,
then assign parent index to largest variable
if(r <= heapSize && A[r] > A[largest]) //compare value of
right child with value of largest index  value
                              //in the array which either parent
node or left child node value
largest = r;   //if right child's value is greater assign
right child index to the largest variable
if(largest != i)  //if largest index not equal to parent index
then swap the values of largest element
        //with parent node value
{
int temp = A[i];
A[i] = A[largest];
A[largest] = temp;
Heapify(A, largest); //Again recursively check current parent
value with its left and right child  value
                       //till parent node reach to the root node
in the binary tree
}
}
void BuildHeap (int* A)
{
int i;  //index of parent node
heapSize = n - 1;  //Heap size is (n-1) because array index
start from 0
for(i = (n - 1) / 2; i >= 0; i--)
{
Heapify(A, i);  //processing of Heap
}
}
//Heap sort function
void HeapSort (int* A)
{
int i;
BuildHeap(A); //Creating Heap function

for(i = n - 1; i > 0; i--)
{
//swap the root and last element into the heap
```

```
int temp = A[heapSize]; //copy last element in array into temp
variable
A[heapSize] = A[0];    //copy root element into last array
position which is equal to heap size
A[0] = temp;           //copy temp into root node variable
heapSize-- ;          //decrease the size of heap by one element
Heapify(A, 0);
}
}
//main function definition
int main()
{
int array[LENGTH];
int i;
printf("Enter the size of array: \n");
scanf("%d", &n);
printf("Enter %d elements in array: \n", n);
for(i=0; i< n; i++)
scanf("%d", & array[i]);
printf("Elements before sorting: \n ");
for(i = 0; i < n; i++)
{
printf("%d\t", array[i]);
}
HeapSort(array);
printf("\nElements after sorting:  \n");
for(i = 0; i < n; i++)
{
printf("%d\t", array[i]);
}
return 0;
}
```

Output:

```
Enter the size of array:
5
Enter 5 elements in array:
76
97
51
21
31
Elements before sorting:
 76      97      51      21      31
Elements after sorting:
21      31      51      76      97
```

7.9.6.6 Analysis of Heap Sort

1. First, consider the creation phase, in this phase, insertion of the key requires O ($\log_2 i$) comparisons and interchange with a maximum of k other keys along the branch to the root of the heap.
2. If we analyze the processing phase, a heap of size i requires O($\log_2 i$) comparisons and interchanges even in the worst case.

Therefore, the required number of comparisons and interchanges on average is,

$$1/2 \sum_{i=2 \text{ to } n} \log_2 i + 1/2 \sum_{i=2 \text{ to } n} \log_2 i = O\left(n \log_2 n\right)$$

Thus, in the worst case, time complexity for heap sort is **O (n $\log_2$ n).**

Time complexity of heap sort:

1. Best case=Ω (n log n)
2. Average case=Θ (n log n)
3. Worst case=O (n log n)

Space complexity of heap sort:

O (1) is the space complexity of heap sort.

7.9.6.7 Advantages of Heap Sort

1. The heap sort is very efficient. Heap sort is particularly suitable for sorting a huge list of items.
2. The heap sort can be implemented as an in-place sorting algorithm. This means that its memory usage is minimal.
3. The heap sort gives us consistent performance. This means it performs equally well in the best, average and worst cases.
4. The heap sort is simpler to understand than other equally efficient sorting algorithms because it does not use recursion.

7.9.6.8 Disadvantages of Heap Sort

1. A stable sort maintains the relative order of items that have the same key that is the way they are present in the initial array. Heap sort is unstable sort. It might rearrange the relative order.
2. In practice, heap sort is slightly worse than quick sort.

7.10 Interview Questions

1. Write and explain a C program to implement a linear search algorithm?
2. Can you implement a binary search algorithm without recursion?
3. Implement and write the bubble sort algorithm?
4. Write advantages and disadvantages of the bubble sort with performance analysis of the Bubble sort?
5. Explain how selection sort works with advantages?
6. Why Sorting algorithms are important?
7. Explain how insertion sort works with performance analysis of insertion sort?
8. What are the advantages and disadvantages of quick sort?
9. Explain and write how heap sort works with a suitable algorithm?
10. Define merge sort. What are the advantages of merge sort?
11. Differentiate linear search and binary search.
12. Explain the linear search algorithm with an example.
13. Write down the merge sort algorithm and give its worst-case, best-case and average-case analysis.
14. Briefly differentiate linear search algorithm from a binary search algorithm.
15. List all types of sorting techniques? Give the advantage of merge sort?
16. What is the difference between searching and sorting?
17. Explain in detail about sorting and different types of sorting techniques.
18. Write a program to explain bubble sort. What is the worst-case and best-case time complexity of bubble sort?
19. Write a C-program for sorting integers in ascending order using insertion sort.
20. Explain the selection sort? Demonstrate the selection sort results for each pass for the following initial array of elements 21, 36, 83, 57, 31, 79, 31, 18 and 55.
21. Explain the algorithm for selection sort and give a suitable example.
22. Explain the analysis of searching techniques with best, average and worst case.
23. Write and explain the C program for binary search with time complexity?

24. Explain the algorithm for quick sort and give a suitable example.
25. Write a program to implement quick sort.
26. What are the different types of hashing techniques explain with a suitable example?
27. What is hashing? Explain collision in hash function with suitable example?
28. Write a note on types of hash functions?
29. What is open and closed hashing?
30. Discuss collision resolution techniques with examples.
31. Write a note on linear probing and chaining without replacement?

7.11 Multiple Choice Questions

1. What is the best case time complexity for insertion sort to sort an array of n elements?

   A. Ω (n)

   B. Ω (n log n)

   C. Ω (n^2)

   D. Ω (n log n^2)

   Answer: (A)
   Explanation:
   In insertion sort, the best case occurs when the array is already in sorted form. In this case, run time control will not enter into the loop.
   Then, **T (n) = 1 + 1 + ... + 1 (n times)**
   T (n) = Ω (n) , so the time complexity of insertion sort remains Ω (n) for the best case.

3. If the number of elements to be sorted is very less, then ...... sorting can be efficient.

   A. Heap sort

   B. Quick sort

   C. Merge sort

   D. Selection sort

   Answer: (D)
   Explanation:
   Selection sort is a good sorting technique when a number of elements are less in the array.

4. Which sorting algorithm is of priority queue sorting technique?
   A. Quick sort
   B. Heap sort
   C. Insertion sort
   D. Selection sort

   Answer: (D)
   Explanation:
   The selection sort consists entirely of a selection phase in which the smallest element, in the array is searched. Once the smallest element is found, it is placed in the first position of the array which is similar to the priority queue sorting technique.
5. Sorting is also known as a partition and exchange type of sort.
   A. Quick sort
   B. Heap sort
   C. Insertion sort
   D. Selection sort

   Answer: (A)
   Explanation:
   Quick sort is known as partition and exchange type of sort.
6. Which of the following sorting algorithm is based on the divide-and-conquer technique?
   A. Bubble sort
   B. Selection sort
   C. Merge sort
   D. Insertion sort

   Answer: (C)
   Explanation:
   Merge sort and quick sort algorithm is based on the divide-and-conquer technique.
7. What is an external sorting algorithm?
   A. Algorithm that uses hard disk while sorting the elements.
   B. Algorithm that uses main memory or primary memory during the sorting of the elements.
   C. Algorithm that involves interchanging of elements
   D. None of the above

   Answer: (A)
   Explanation:
   External sorting is sort of elements from an external file by reading it from secondary memory.

8. The time complexity analysis of the heap sort in the worst-case scenario is

   A. O(n2 log n)

   B. O(log n)

   C. O(n log n)

   D. $O(n^2 \log n^2)$

   Answer: (C)
   Explanation:
   The worst-case time complexity of heap sort is O (n log n).

9. If the given input array elements are almost sorted in this scenario, which of the following internal sorting algorithm gives the optimum performance efficiency?

   A. Selection sort

   B. Bubble sort

   C. Heap sort

   D. Insertion sort

   Answer: (D)
   Explanation:
   If the given input array elements are almost sorted then the insertion sort algorithm gives you Ω (n) time complexity.

10. What is special algorithm design paradigm is used in the quick sort and merge sort algorithm?

    A. Greedy method paradigm

    B. Backtracking paradigm

    C. Divide-and-conquer paradigm

    D. Dynamic programming paradigm

    Answer: (C)
    Explanation:
    The quick sort and merge sort algorithms are based on the divide-and-conquer design paradigm.

11. What is the complexity of the search time of the hashing search method?

    A. O (n2)

    B. O (logn)

    C. O (nlogn)

    D. O (1)

    Answer: (D)
    Explanation:

Since every hash key has a unique array index, requires a constant time for searching an element from the array.

12. Which of the following is not the technique to avoid collision in hashing?

    A. Linear probing

    B. Chaining with replacement

    C. Chaining without replacement

    D. Dynamic programming

    Answer: (D)
    Explanation:
    Linear probing, chaining with replacement and chaining without replacement are the techniques to avoid collision in hashing. While dynamic programming is not the collision avoiding technique.

13. In the chaining technique in hashing, which of the following data structure is most suitable?

    A. Singly linear linked list

    B. Doubly linked list

    C. Tree

    D. Graph

    Answer: (B)
    Explanation:
    In a doubly linked list, deletion of elements from chaining becomes easier than other data structures, hence a doubly linked list is most suitable.

14. What is the worst-case time complexity of linear search and binary search, respectively?

    A. O (n log2 n), O(log2 n)

    B. O(log2 n), O(n)

    C. O(n), O(log2 n)

    D. O(1), O(n)

    Answer: (C)
    Explanation:
    The worst-case time complexity of the linear search is O (n) and for binary search is O ($\log_2$ n).

15. The case in which a hash key other than the desired one is kept at the identified location is called as?

    A. Open hashing

    B. Closed hashing

C. Chaining

D. Collision

Answer: (D)
Explanation:
When a hash function returns the same address or hash key for more than one record is called a collision.

16. Which of the following data structure is used in hash tables?

A. Queue

B. Doubly linked list

C. Stack

D. Array

Answer: (B)
Explanation:
The linked list data structure is used to organize the data in hash tables.

17. What is the time complexity of searching, deleting and inserting into direct addressing in hashing?

A. O(1), O(n), O(1)

B. O(n), O(n), O(1)

C. O(1), O(1), O(1)

D. O(1), O(n), O(n2)

Answer: (C)
Explanation:
Hashing functions such as inserting, deleting and searching operations can be performed with a constant time complexity equal to O (1).

18. Which of the following sorting technique is called non-linear sorting?

A. **Heap sort**

B. Bubble sort

C. Insertion sort

D. Quick sort

Answer: (A)
Explanation:
Heap sort is also called a non-linear sorting technique.

19. Which of the following is not an example of a closed hashing or open addressing method?

A. Linear probing

B. Quadratic probing

C. Double Hashing

D. Chaining

Answer: (D)
Explanation:
Chaining is an example of open hashing or closed addressing method.

20. What is the quickest search method among the following in the worst case?
    A. Linear searching
    B. Binary searching
    C. Hashing
    D. Sequential search

    Answer: (C)
    Explanation:
    Hashing is the fastest search method as it gives direct addressing also in the worst case.

21. Which of the following is an example of an open hashing or closed addressing method?
    A. Linear probing
    B. Quadratic probing
    C. Double Hashing
    D. Use of buckets

    Answer: (D)
    Explanation:
    The use of buckets is an example of open hashing or closed addressing method.

22. What is the space complexity of quick sort and merge sort algorithm?
    A. O (n log2 n), O (log2 n)
    B. O (log2 n), O (n)
    C. O (n), O (log2 n)
    D. O (1), O (n)

    Answer: (B)
    Explanation:
    The space complexity of quick sort is O ($\log_2$ n) and the merge sort algorithm having O (n) as space complexity.

References

Alfred V. Aho, John E. Hopcroft, and Jeffrey D. Ullman 2012. *Data Structures and Algorithms*. Published by Pearson Education, Bengaluru.

Ellis Horowitz, Sartaj Sahni, and Susan Anderson-Freed 1992. *Fundamentals of Data Structures in C*. Published by W. H. Freeman, Macmillan Publishers, New York.

G. S. Baluja 2014. *Data Structures through C++ (A Practical Approach)*. Published by Dhanpat Rai & Co. (P) Limited, New Delhi.

Kamal Rawat, Meenakshi 2018. *Data Structures for Coding Interviews*. Published by BPB Publication, Delhi.

Narasimha Karumanchi 2016. *Data Structures and Algorithms Made Easy: Data Structures and Algorithmic Puzzles*. Published by CareerMonk, Hyderabad.

Peter Brass 2008. *Advanced data structures*. Published by Cambridge University Press, Cambridge.

Seymour Lipschutz 2014. *Data Structures, (Schaum's Outline Series)*. Published by McGraw Hill Education, New York.

Thomas H. Cormen, Charles E. Leiserson, Ronald L. Rivest, and Clifford Stein 2014. *Introduction to Algorithms*. Published by The MIT Press, Cambridge, MA; London.

Vibrant Publishers 2016. *Data Structures & Algorithms Interview Questions You'll Most Likely Be Asked*. Published by Vibrant Publishers, Chennai.

Y. Langsam, M. J. Augenstein, and A. M. Tenenbaum 2015. *Data Structures Using C and C+*. Published by Pearson Education, Bengaluru.

Index

Printed in the United States
by Baker & Taylor Publisher Services